CARMONTELLE
GARDEN AT MONCEAU

CARMONTELLE

GARDEN AT MONCEAU

Elizabeth Barlow Rogers, *Editor*

Joseph Disponzio, *Managing Editor*

Translated by Andrew Ayers

With contributions from:
Laurence Chatel de Brancion
Joseph Disponzio
Florence Gétreau
David L. Hays
Elizabeth Hyde
Susan Taylor-Leduc
Caroline Weber
Gabriel Wick

Foundation for Landscape Studies · NEW YORK, NY
Oak Spring Garden Foundation · UPPERVILLE, VA
Distributed by Yale University Press · NEW HAVEN & LONDON

Published by
The Foundation for Landscape Studies, New York, NY
The Oak Spring Garden Foundation, Upperville, VA
Distributed by Yale University Press
302 Temple Street
P.O. Box 209040
New Haven, CT 06520-9040
yalebooks.com/art

ISBN 978-0-300-25468-6
Library of Congress Control Number: 2020935673

FRONTISPIECE: Carmontelle, *Self-portrait*, ca. 1765.

10 9 8 7 6 5 4 3 2 1
PRINTED IN CHINA

Contents

Preface *Elizabeth Barlow Rogers and Peter Crane*   VII

Introduction *Joseph Disponzio*   I

GARDEN AT MONCEAU

Translator's Foreword *Andrew Ayers*   5

Carmontelle · Prospectus for the *Garden at Monceau*   7

Carmontelle · *Garden at Monceau*   11

Carmontelle and his World *Laurence Chatel de Brancion*   63

ESSAYS

History by Design: The Aesthetics of Transformation in Carmontelle's
Jardin de Monceau
  David L. Hays   97

Fashion Follies at the *folie de Chartres*
  Caroline Weber   105

Carmontelle's Portraits of Musicians
  Florence Gétreau   111

"The Habit of Seeing the Same Things Often":
Planting the Picturesque in Eighteenth-Century France
  Elizabeth Hyde   127

Monceau, the *Mémoires secrets*, and the Colisée: The Reinvention of
the duc de Chartres *à l'anglois*
  Gabriel Wick   135

A Spade in the Garden or a Garden in Spades: Louis Carrogis de Carmontelle's
Jardin de Monceau or the Garden as Dramatic Proverb
  Joseph Disponzio   141

En-jeux: Viewing, Mapping, and Playing in Carmontelle's Prospectus and
Jardin de Monceau
  Susan Taylor-Leduc   153

Notes   163

Image Credits   180

Contributors   183

Preface

Landscape design history is an expression of cultural values, stylistic influences, and individual creativity. This book grew out of collaboration between the Foundation for Landscape Studies and the Oak Spring Library at the Oak Spring Garden Foundation to publish one of the key texts in French Picturesque garden history, *Jardin de Monceau* (Garden at Monceau, 1799) by Louis Carrogis, known as Carmontelle (1717–1806). The project reflects a mutual and complementary goal of our respective organizations: to promote scholarship in garden history and landscape studies through the English-language translation and facsimile reproduction of certain period volumes in rare-book library collections. In this instance, the chosen folio is a particularly fine copy of Carmontelle's text in the Oak Spring Library amassed by Rachel Lambert Mellon (1910–2014).

Our purpose in publishing the current edition of *Garden at Monceau* is two-fold: to make a valuable but little-known book accessible to landscape historians, and to open for all readers a window on the life of the ancien régime on the eve of the French Revolution. In addition to the translation of Carmontelle's text and the prospectus that preceded it and the high-quality reproductions of the original eighteen engraved plates, this edition contains eight expository essays by noted landscape, cultural, and social historians. These scholars illuminate the forces at play in the fantasyland Carmontelle created for the libertine duc de Chartres, the future Philippe-Égalité. They also explore Carmontelle's eccentric interpretation of French Picturesque garden theory.

We thank the board of the Oak Spring Garden Foundation and donors to the Foundation for Landscape Studies, whose support has made this project possible. In addition, we offer our gratitude to managing editor Joseph Disponzio and the scholars with whom he has collaborated to create this work. Special recognition goes to William Christie, founding director of Les Arts Florissants, whose advisory role has been of immeasurable value. We also acknowledge with appreciation the museums, institutions, libraries, and private collectors who have allowed for the reproductions of illustrations, without which this publication would have been materially diminished.

Elizabeth Barlow Rogers, *President, Foundation for Landscape Studies*
Sir Peter Crane, *President, Oak Spring Garden Foundation*

INTRODUCTION

Joseph Disponzio

CARMONTELLE's folio *Jardin de Monceau* and its prospectus have long been staples in the study of the French Picturesque garden. Yet, despite the renown of Carmontelle and the garden itself, until recently the works have been hard to come by. While Carmontelle's folio is now available on the Gallica platform of the Bibliothèque nationale de France, the oft-cited and important prospectus is not. Neither has ever appeared in an easily accessible, contemporary French edition, let alone in English.[1] Thus the publication of Carmontelle's two texts in Andrew Ayers's unabridged English translation is a welcome event in garden-history scholarship.

The major English texts on Picturesque garden theory, including Thomas Whately's 1770 *Observations on Modern Gardening* and William Chambers's 1772 *Dissertation on Oriental Gardening*, are now easily accessible to modern readers, and with the translation of the *Jardin de Monceau*, the major French texts—these include Claude-Henri Watelet's 1774 *Essai sur les Jardins*, Jean-Marie Morel's 1776 *Théorie des jardins*, and René de Girardin's 1777 *De la composition des paysage*—are all available in English.[2]

These texts constitute the range of theoretical approaches to the Picturesque or natural garden, which was initiated by the English in the early- to mid-1700s and quickly imitated by the French. At one end of the theoretical continuum were partisans of the landscape style espoused by Capability Brown, in which long vistas, broad lawns, clumps of trees, and bodies of water predominated, and architectural follies or other structures were few. In this camp could be placed Whately, Morel, Girardin, and Watelet. In broad terms, one could say that the Brownian school favored nature over art—that is, its adherents trusted that the manipulation of nature alone could create landscape gardens of compelling interest and sentiment. At the other end was William Chambers, who believed that nature left to herself could not move the soul; the artful intervention of human agency was needed. Where Brown's designs were naturalistic, Chambers opted for artifice.

Centrally located in the Chambers camp was Carmontelle. He was a masterful, one-of-a-kind entertainer-artist-writer whom perhaps only the ancien régime could have produced. Attached to a princely household, he was responsible above all for amusing the court. When the craze for English gardens hit Paris, Carmontelle was quick to exploit the new garden style, producing one of the most extravagant gardens of pre-Revolutionary France. To say that his garden at Monceau is sui generis doesn't capture its uniqueness. It was less a garden and more an amusement park—a *parc d'attraction* in contemporary French parlance, something akin to Disney theme parks.

To promote (as well as memorialize) his garden, Carmonelle produced his sumptuous folio: a broadsheet-sized brochure containing the text, accompanied by a plan of the garden and illustrations. The descriptive text identified the constituent parts of the garden, although with creative embellishments that were suggestive of a fictive narrative. Importantly, the text presented Carmontelle's theoretical approach to the Picturesque garden—an approach that stood in stark contrast to that of his countrymen.

With characteristic sarcasm, Carmontelle delivered his synopsis in the form of pithy punch lines and humorous retorts. In a dozen or so pages of sometimes erratic prose, he said what he had to say about his gardening approach, his brevity serving as a sly rebuke to the hundreds of pages his contemporaries needed to get their points across.

But it is the plates, more than the text, that have made Carmontelle's folio justly famous and expensive.[3] Taken as a whole, the suite of illustrations constitutes a stunning example of the art of engraving in the eighteenth century, executed at the highest level. These plates are why Carmontelle's Monceau has held the interest and attention of connoisseurs, scholars, and collectors for centuries. As works of art they speak for themselves, but they also function as vignettes: they hint at stories and conjure flights of fancy of limitless interpretation. Those reproduced here are from the library of the Oak Spring Garden Foundation.

To help the reader appreciate the multiple meanings and associations embedded in Carmontelle's Monceau, we have invited several noted scholars to guide the way. Eminent Carmontelle historian Laurence Chatel de Brancion gives us an atmospheric and insightful view of Carmontelle and his world, and of the creation of Monceau and its aftermath. David L. Hays, whose extensive scholarship on Carmontelle is fundamental to the field of French garden history, further contributes to his seminal work on Monceau by disinterring the impossibly complex acquisition history of garden's site and relating it to the design history of the garden. Caroline Weber, a professor of French, and Florence Gétreau, a musicologist, focus on social representation in Carmontelle's Monceau. The former analyzes hairstyles and dress; the latter spotlights music and musicians. Elizabeth Hyde provides a horticultural look at Monceau that other scholars have been slow to exploit. Gabriel Wick explores the political interpretations to which the design of Monceau lends itself, while Susan Taylor-Leduc examines the garden in terms of game playing. My own essay focuses on garden elements that seem comical and even frivolous, but when viewed through a different lens prove to be nothing of the sort. The multiple readings of the garden presented in the essays bear witness to the fecundity of Carmontelle's creativity, wit, and inventiveness.

Robert Harbison has written that "gardens always mean something else, man absolutely uses one thing to say another."[4] He could have been speaking of Monceau and its creator, so much is a normative analysis of the garden impossible: its meaning is open-ended. Yet it is precisely because Monceau was so beguiling that it still captures the imagination of anyone willing to be drawn into Carmontelle's world. Few works of the gardening art can make such a claim.

1. Carmontelle's *Jardin de Monceau* was published in a limited, facsimile edition by Jardin de Flore in Paris in 1979, which is available at considerable cost.

2. These are but a few of the key texts published during the seminal decade of 1770s. Whately is available in a modern edition, with an introduction and commentary by Michael Symes (Rochester, NY: Boydell Press, 2016). The second revised edition of Chambers (London: Printed by W. Griffin, 1773) is available online at https://archive.org/details/dissertationonor-oocham/page/n7, accessed November 5, 2019. Modern English translations of the French treatises include Claude-Henri Watelet, *Essay on Gardens*, trans. Samuel Danon, intro. Joseph Disponzio (Philadelphia: University of Pennsylvania Press, 2001) and Jean-Marie Morel, *Theory of Gardens*, trans. Emily Cooperman, intro. Joseph Disponzio (Washington, DC: Dumbarton Oaks Research Library and Collection, 2019). René de Girardin's treatise was published in English in the eighteenth century as *An Essay on Landscape* (London: J. Dodsley, 1783). This early translation, attributed variously to Daniel Malthus or Jane Dalton, remains the only English version of Girardin's work. It was reprinted most recently by Garland Press (New York and London, 1982), with an introduction by John Dixon Hunt.

3. A complete folio of Carmontelle's *Jardin de Monceau* sells for tens of thousands of dollars.

4. Robert Harbison, *Eccentric Spaces* (Cambridge, MA: MIT Press, 1977), 20.

GARDEN AT MONCEAU

Translator's Foreword

❧

CARMONTELLE'S 1779 *Jardin de Monceau*, which was preceded in December 1778 by a subscribers' prospectus intended to raise funds to cover its printing, belongs to a literary genre all of its own. Certainly it cannot be compared to any of the discursive, philosophical treatises on the theory of garden design that its author professes to admire (Whately's, d'Harcourt's, and Chambers's), nor to the one he attacks but does not name (Morel's). In Carmontelle's text, aesthetic and theoretical discourse fills barely four pages, the overwhelming majority of the work consisting in a detailed description of Monceau, primarily in the form of commentaries accompanying seventeen engravings of its principal views. The exercise was clearly intended to preserve for posterity a creation that time and fashion were threatening to destroy.

In his introductory paragraphs, Carmontelle flits from subject to subject like the salon habitué he was, never dwelling too long on any (it wouldn't do to bore one's reader like that old pedant Morel) in what appears to be a list of points and score settling intended for an audience whose members he all knew. As in a roman à clef—an aristocratic genre par excellence during the ancien régime—his target readership would have recognized exactly to whom he was referring with every gripe and prick of persiflage. The tone and vocabulary are conversational, with none of the learned ponderousness that French authors are sometimes wont to adopt.

This translation attempts to reflect that vocabulary and conversational tone, while following as closely as possible Carmontelle's syntax and word order; there has been no attempt to adjust its rhythms to the modern ear. Nor, on the other hand, has there been any endeavor to use English archaisms. To keep a flavor of the original, some of the typesetting quirks have been reproduced—the rather random use of initial capitals for nouns, the use of ampersands, the italicization, the distinctive spacing of the letters in titles—but the spelling is modern, except for proper nouns. These have been left in their original forms, with footnotes indicating the modern orthography and, in the case of individuals, dates of birth. Other translator's notes explain particular points where necessary. Punctuation follows modern usage, but Carmontelle's rather random chapter divisions have, of course, been retained.

ANDREW AYERS

December 2019

GARDEN
AT MONCEAU,
NEAR PARIS,

Belonging to H. M. S. H. Monseigneur the Duke OF CHARTRES.

THIS GARDEN having aroused the curiosity of a lot of people, & even Foreigners, many are those who initially judged it to its disadvantage, by seeing it from the exterior: on visiting it, they recognized that seeing the features in proportioned distances produced a better effect & made them seem larger, since they were no longer compared with the surrounding immensity. When people thought there were too many things in this Garden, it was because they couldn't be bothered to wait until the trees had grown to form masses & divide the terrain; people still judge the Garden this way in winter, on seeing all the features of which it is composed one upon the other, like Engravings viewed together through the paper. This sort of Garden can therefore not be judged like ordinary Gardens. If one did not know what the Gardens of Versailles were like before they were replanted, would one believe that they are Le Nostre's[1] masterpiece, seeing them as they are now, where all that adorns them can be seen at once? Everything seems very close together, & it is difficult to comprehend their beauty & vast extent.

We did not wish to create an English Garden at Monceau but exactly what we said when critiquing English Gardens: to unite in one single Garden *all times & all places*. It is a simple fantasy, the desire to create an extraordinary Garden, a pure amusement, & not the desire to imitate a Nation which, when making natural Gardens, runs the roller over all the lawns & spoils nature by showing everywhere the prissy art of the unimaginative Gardener, as says *Mr. Chambert* in his preface to the *Dissertation on Oriental Gardening*.[2] However he cites some that are beyond criticism; but those are not the ones that are imitated when one wishes to create an English Garden.

We are not unaware that, to create a natural Garden, one should avoid choosing a plain where there is not a single shrub nor drop of water. We know that one should select land on which there are hills, ponds, streams, meadows, woods, rocks, &c. With knowledge of the picturesque, with taste, & with imagination, one may create a very handsome Garden, if one finds old trees that have been planted & which, on spreading their branches, cast shade on some parts & render brighter those where the light dominates. If one has water that one can capture & collect, one can make rivers out of it that will flow agreeably without being held in by cement-coated walls. As

---

1. André Le Nôtre (1613–1700).
2. William Chambers (1723–96). *A Dissertation on Oriental Gardening* was published in 1772.

a result, they will easily maintain the freshness & greenness of the meadows. In the woods, one will create clearings, & mix evergreen & foreign trees, which will vary the color of the leaves & of the shrubs by producing flowers of different hues over several seasons, & by following the art of creating modern Gardens, by the Englishman *Sir Thomas Wateley*,[3] one will produce a very handsome Garden, especially by never forgetting to contemplate[4] nature's most beautiful effects, which one should study as a Painter, by exploring them in the countryside, pencil in hand.

In the Garden at Monceau, everything had to be created; we took advantage only of the views of the surroundings, which we appropriated by enclosing the Garden with nothing but ditches, which makes it impossible to estimate its true extent.

A Garden near to Paris, where one wishes to spend only a few hours, is big enough at forty-five arpents,[5] since it will take nearly two hours to walk through it without stopping. If the main features were further away from each other, one would have to visit it on horseback or in a carriage. Had we included fewer features, there would have been less variety, & the different scenes that appear at each moment are the source of all the pleasure.

In any event, it is these scenes that we are currently offering to the Public. *Mr. de Lafosse*, following the drawings of *Mr. de Carmontelle*, the Author of this Garden, has undertaken this enterprise, & one may surely count on the care that he is able to bring to it.

A general plan will set forth the entire Garden, & seventeen views will show its principal effects; meaning there will be a total of eighteen plates. These views will be named, & a point marked on the plan by a letter will indicate the place from where the picture was drawn.

> The first plate will be the overall plan.
> The second, the view of the water mill & the bridge leading to it, above which Montmartre can be seen.
> The third, the view of the entrance to the Garden & the principal pavilion.
> The fourth, the view of the other side of the water mill bridge, & of the ruins of a gothic castle, on which water falls in a cascade.
> The fifth, the view of the farm.
> The sixth, the view of the ruins of the Temple of Mars.
> The seventh, the view of the Isle of Rocks, the Dutch windmill, & the rocks[6] which are the source of all the water in the Garden.
> The eighth, the view of the Minaret hillock.
> The ninth, the view of the Italian vineyard.
> The tenth, the view of the wooden bridge.
> The eleventh, the view of the Circus or Naumachia.
> The twelfth, the view of the wood of tombs.
> The thirteenth, the view of the two pavilions.
> The fourteenth, the view of the Tartar tent.
> The fifteenth, the view of the white-marble temple.

---

3. Sir Thomas Whately (1726–72). The reference is to Whately's *Observations on Modern Gardening*, published in 1770.
4. Carmontelle uses the verb "se rapprocher de," literally, "to draw near to."
5. The *arpent* was an eighteenth-century land measure based on the *perche*. It had no uniform standard and varied from locale to locale. For a detailed discussion of the size of Monceau, see David L. Hays's contribution to this volume.
6. See footnote 13, page 16.

The sixteenth, the view of the Turkish tents.

The seventeenth, the main view[7] of the pavilion & the carousel.

The eighteenth, the view of the grove of the sleeping Faun.

All these views are embellished with figures, & each print will be sold for thirty sols, & will measure seventeen *pouces* in length, by ten & a quarter high (1). There will be an explanation of each print. The full price of this Work will be twenty-seven livres for Subscribers & forty livres for those who have not subscribed. There will be three deliveries of six prints each with their text: nine livres must be paid on receipt, & the impressions will follow the order of the date of Subscription, & the list of Subscribers will be printed following this order. One need only sign the Subscription slip.

Subscriptions may be made until the first of April 1779 *at Mr. D E   L A F O S S E, Engraver by Royal Appointment, rue du Carrousel, near the Tuileries; at Messrs. N É E & M A S Q U E L I E R, rue des Francs-Bourgeois, Porte St. Michel; & at every Vendor of Prints.*

From the day on which this *Prospectus* is published, only postage-paid letters will be accepted.

---

(1) Their size makes them most suitable for viewing in peepshow boxes.

Read & approved this December 23, 1778, DE SAUVIGNY.

Whereas, approval has been given, license to print is granted, subject to recording at the Trade Chamber, December 24, 1778, LE NOIR.

The present license has been recorded on the Police Register of the Royal Trade Chamber of Booksellers & Printers of Paris, no. 7015, in accordance with the old Regulations confirmed by that of February 28, 1723. Paris, this December 29, 1778,

A. M. L O T T I N the Elder, member of the Chamber.

From the Printing House of L. J O R R Y, rue de la Huchette.

To Mr.,
Mr. D E L A F O S S E, Engraver, rue & place du Carrousel, opposite the Porte des Tuileries,
P A R I S.

I undertake to purchase
Example   *of the* Picturesque Views of the Garden at Monceau, *of which the person charged with the distribution of this Work will successively deliver to me the different impressions.*

In   , this   1779.

My address is

---

7. A typesetting error? The French reads "vue principale du pavillon"; elsewhere Carmontelle calls it the "pavillon principal."

JARDIN

DE

MONCEAU,

PRÈS DE PARIS,

APPARTENANT

A SON ALTESSE SÉRÉNISSIME

MONSEIGNEUR

LE DUC DE CHARTRES.

A PARIS,

Chez {
M. DELAFOSSE, Graveur, rue du Carrousel, près des Tuileries;
MM. NÉE & MASQUELIER, Graveurs, rue des Francs - Bourgeois, vis - à - vis de la rue de Vaugirard;
}

Et chez tous les Marchands d'Estampes.

M. DCC. LXXIX.

GARDEN

AT

MONCEAU,

NEAR PARIS,

BELONGING

TO HIS MOST SERENE HIGHNESS

MONSEIGNEUR

THE DUKE OF CHARTRES.

IN PARIS,

Available from { Mr. Delafosse, Engraver, rue du Carrousel, near the Tuileries;
Messrs. Née & Masquelier, Engravers, rue des Francs-Bourgeois, opposite the
rue de Vaugirard;

M. DCC. LXXIX.

(ii)

FOREWORD.

*WE do not have the pretension of offering here a theory or precepts;
it would be ridiculous, knowing the work published in England by Sir
Thomas Wathely[1] under the modest title of* Observations on Modern
Gardening. *This Book contains the true elements of these sorts of Gardens. Other works of this type, written by French Connoisseurs, are useful
& agreeable; but there is one, little known, so filled with taste & imagination, that we must truly regret that the Author (I) has not had it printed.*

*The Garden at Monceau is therefore not a model that we suggest be
followed. The goal of perfection is so far off, that one must study & practice for a long time before even daring to hope to reach it.*

*We have allowed ourselves here a few reflections on new Gardens, &
the observation that, our fortunes, our morals, our tastes, & our climate
being different from those of the English, our Gardens must not be servile
imitations of theirs, but composed in light of these differences.*

(1) The D. D'h....[2]

1. An approximation of Whately's family name.
2. François-Henri d'Harcourt, fifth duc d'Harcourt (1726–1802), whose *Traité de la décoration des dehors, des jardins et des parcs* was written
   in 1775 but not published until 1919 (by Ernest de Ganay).

GARDEN

AT

MONCEAU.

CHAPTER ONE.

THE goal of the Arts in great things is to force the soul into admiration, to carry off, bind, & dominate it through the power that real beauty always has the right to exercise over them.[3] In those that are but agreeable, the soul must continually be moved, interested, & amused by the charm of images, because it cannot do without them; but is everyone sensitive to the products of the Arts? Without practicing them, do we have the sure & developed taste that is so necessary to appreciate them well? For that, we would need to converse often with Artists; only they can make us see how ridiculous it is to judge by the word of others, & that with constant & reflective study we can bring ourselves to a position where we can judge for ourselves. Thus, in order to sense well, we must acquire the talent of seeing well.

We have masters to help us learn to talk, dance, & sing, &c., & yet we do not think of learning to see. How many pleasures do we deprive ourselves of by neglecting this science! It is with Painters that we may acquire it. Stroll with them through nature, & they will stop you at every step in order to have you observe its beauties. They shed light on everything you see, they make you perceive the gradations of linear & aerial perspective, they unveil the secrets of space; they will reveal to you the diversity of all the color tones, their relationships, & their harmony. Without such knowledge, we see only the skeleton of nature; for that reason children & the ignorant want to touch all that they see, & it is at their fingertips that they have their eyes.

For the man who makes us see, everything in nature is a spectacle, & the slightest detailed object becomes for him a source of reflection.

We would embarrass many of those who pride themselves on their love for the countryside if we forced them to describe exactly what it is that makes them like it so. They would perhaps be reduced to repeating the poetic descriptions they have read of it & which they have never compared with nature. Yet they claim to enjoy all its charms. Eh! what are the country charms that are cited in France? Pure air & freedom; & one almost never enjoys either one or the other. But would they be sufficient for us? We would find few charms in the countryside without those of society; it is there that we truly enjoy them, where we get to know each other more in less time than we would have done during the winter had we supped together every day. A Frenchman cannot get by with just his own company; the desire to charm, to love, & be loved are for him needs; he never separates his existence from that of others, however high his opinion of himself. It is therefore in the countryside that we best taste the sweetness of being together, that we get to know each other

---

3. "Them" being "great things" ("elles" in the original French).

better, that we choose each other, & that liaisons form which constantly renew the pleasures of being there; talents develop there; & if, in a region, there are several households that share the same tastes, when they get together, with no special preparation, everything becomes festive.

We need our pleasures in the countryside. Despite the charms that nature can offer, we require good food, hunting, gaming, concerts, shows—that is what we desire there & what we praise there. That does not mean that we banish philosophical conversation from the countryside; but our philosophy there is not austere, dark, & untamed. Instead of avoiding each other to go off & dream, we seek each other out to talk; in the countryside, we are far removed from any desire to feed melancholy; we talk politics, but no one believes they are governing. We make it our business to please the ladies—it is they who are the delight of society. Far, therefore, from abandoning them to themselves, as do the English, we do only that which suits them; but it is difficult to persuade them to go for a walk, & it is always late when they go out: then, deprived of bright light & having less brilliance, features lose much of their charm, & the dampness of the woods & the meadows, as well as persecution by insects, bring the walker back to the main paths. That is how we enjoy the charms of nature, which is to say we barely even think of them. But what does it matter?—we talked, we laughed, we were gay.

If sometimes we make a resolution to see the surroundings in the place where we are staying, then the whole company sets off, & those it comprises climb to the high spots; they count the bell towers & judge the view only by the remoteness of the objects they perceive. The gradation between those that are near to them & those that are further away does not interest them; they all run after the spyglass so as to have the pleasure of working out what is in the vapors of the air & arguing over the forms & names of the objects. What they call a fine view is but a horizon of which they see only the points where it terminates; & if one wants to explain to them what produces a handsome picture, they are astonished that one would want to teach it to them, they do not listen to you, or are distracted; yet, without knowing anything more, they judge none the less, when the occasion arises, both the work & the Artist; because it is customary & because he who is the least knowledgeable is always the one who judges & decides with the greatest confidence.

The habit of seeing the same things often makes us ignore their value. How many people have long walked in the Tuileries without suspecting this Garden's beauty! And how many others have crossed the Louvre without knowing that its Peristyle[4] is one of the most handsome monuments of Architecture that exists in Europe, while all the Foreigners are struck by its grandeur & magnificence!

It is therefore not enough, in order to hold the attention, to reproduce the known: one must also use new, varied, & unforeseen means; & when one is lucky enough that he to whom the Garden belongs makes it a rule that you must follow & steadfastly refuses to have in it what one finds in all the others, then this rule makes you seek out new ideas, combine them, & multiply them.

If one can turn a picturesque Garden into a land of illusions, why refuse oneself the opportunity? Only illusions are amusing; if liberty guides them, may Art direct them, & we will never stray from nature. Nature is varied depending on climate; let us try, through illusory means, to vary climate too, or rather to make ourselves forget the climate in which we find ourselves; let us bring to our Gardens the Scene changes of Operas; let us show in them, in reality, what the cleverest scene painters could propose as decorations, *all times & all places*. Let us be allowed to avoid this cold monotony produced by supposedly severe precepts that restrict the imagination. Since everything must be created, let us use this freedom to enchant, to amuse, & to arouse interest. This is what those who came to see the Garden at Monceau were hoping to find; be-

---

4. The more correct term would be "colonnade," and indeed this is the word generally used in French today. At the time, however, it was common to talk of the "péristyle du Louvre."

cause they said when there was not much, *I would be just as happy to walk in the countryside*. People therefore expected to find in the Garden that which they do not ordinarily see elsewhere. When one is not expecting to find in that which belongs to Princes, the greatest efforts of Art, one hopes to see things that are at least uncommon; & yet often one criticizes what one desired: novelty.

Because the rivers are fabrications in the Garden at Monceau, it was said that there was too much water in the meadows: the bridge railings, they said, should be white; because they are white in England, is that an authority? All of those in Germany are white, red, & black: those of our Gardens would therefore have to be painted the same if the Germans had Gardens that were in fashion. In England there is a mansion built in black stone, with white pointing. Would that be pleasant to imitate? There must be harmony in Gardens, & yellow & red agree better with the greenery of the different trees.

In climates that are hotter than ours, stone takes on red, yellow, & violet tones; & these tones, which are warm, always agree better with the greens of the landscape than those of our stone, which in the countryside remains constantly white. In Germany, all the houses are painted with adornments & figures; those of the Villages in Flanders are built in grey stone & brick, & seen through the foliage of the trees they produce the most joyful & agreeable effect.

When building the Château & the Pavilions at Marly, which are in the middle of the Gardens, we did not hesitate to paint them to resemble marble, to link them more to the greenery, & I believe that we may have confidence in the Organizers of these charming Gardens. The old Trianon always seemed like a Fairy Palace; here are model Buildings for pleasure Gardens. Let us just remember an instant how much, on returning to France, the white houses of our villages make our countryside sad to see, & how much they help to mar it, especially after harvest.

If the fog in England maintains the freshness of the lawns, what point is there in wanting to have the same in France? They are one of the principal beauties of English Gardens; but they cannot be obtained without great expense & without the greatest care. Do we think a Frenchman would constantly look after one? It's a beautiful carpet of green cloth, which deserves the greatest praise for its perfection; & yet Poets will never find the same charm in singing the praises of these lawns as in singing the praises of our flower-strewn meadows; & indeed, can they be compared? Is there anything richer, more agreeable, & more joyful? Picturesque Gardens need fortuitous negligences, which belong only to taste; moreover a swathe of green that is too immense & of the same tone would be too sorrowful for our soul, which desires only sweet, lively, & gay impressions. Let us therefore not believe we need Gardens for thinkers. If we felt the desire to imitate English Gardens, it was only to escape the monotony of ours, & our new Gardens will not always be badly done, because they will not slavishly resemble those of the English. We love this happy freedom that produces new & piquant effects; what is more we have our own ideas, tastes, & customs; if they depend on our climate, any attempt to have us adopt those of our neighbors will be in vain; when we diverge from our old principles, we will invent new ones that will belong only to us: while one may divert the waters of a spring, one cannot alter their quality.

It is said that a Garden can be a country,[5] but one does not create a country; one may embellish it with much talent & taste, as the Lord of Ermenonville[6] has done, who did not stint on his ambitions, who chose his location well, & who has created, of its type, a Garden that is unique in France. Moreover one should be able, like him, to spend one's life enjoying it with one's friends & family.

---

5. The word Carmontelle uses here is "pays," and it is very likely a reference to the 1776 treatise *Théorie des jardins,* written by garden designer Jean-Marie Morel (1728–1810), who named four genres—*parc, pays, jardin proprement dit,* and *ferme*. The "pays"—defined as a large tract of landscape with a natural, rural character—was Morel's favorite genre; thus Carmontelle's criticism is deliberately pointed.
6. René-Louis de Girardin, Marquis of Vauvray (1735–1808).

When people say that one can restrict oneself to a farm,[7] & that a farm can be a Garden, because it can combine the useful & the agreeable, it is very difficult for such a Garden to be suitable for those who, among us, are in a position to create one. We would not be able to enjoy the cares of a farm; its multiple occupations leave no time for leisure, & the details of rural life are ill-suited to our taste for society, pleasure, & dissipation; we prefer the description of them to the practice. Do we think that most of our Ladies, who sit in a Drawing Room whose shutters are tightly closed when the sun shines, would find it charming to follow the labors of a farm the whole day through, to oversee the harvests & the grape gatherings, & to watch the tending of the animals? They find all of that charming in Verse & in Pictures; & things cannot be otherwise, since they only know the inhabitants of the countryside in the most agreeable forms; & ours are so unlike the Shepherds in *L'Astrée*,[8] or those depicted by Fontenelle,[9] or Boucher,[10] or at the Opera, that they[11] will easily destroy in French Ladies the taste they[12] believe they have for purely pastoral life.

We therefore need only Gardens where nature is present in its most agreeable forms; the charm one must feel on entering should be perpetuated, & renewed in every way possible, so as to awaken in the soul the desire to see it every day & to understand it. The true Art is being able to keep the stroller present through the variety of features without which he will go off to find, in the open countryside, what is missing for him in this Garden—the image of freedom.

The pleasure of a natural Garden is to find in it at every step a Scene; & each feature must be disposed so as to produce many, according to the different effects of the light.

The plan of an ordinary Garden allows us to see everything it contains; & once we've seen it, we can dispense with visiting the Garden after that, because we know what clipped hornbeam is; what paths, groves, & pools are, &c. There are thus only the statues left to see, when there are any; because we can easily guess the effects of trellises & waterfalls, which are regular. But in a natural Garden the plan is only an itinerary, & does not provide just a single scene; moreover, the irregularity does not imprint itself on the mind, & one does not foresee the different features that present themselves one after the other, especially when the trees are in full leaf, which is really the only time at which one should see any Garden, if one wishes to judge it.

❦

The plan of the Garden at Monceau [PLATE I] shows the location of the features & the slopes of the terrain; because one can see the flow of the waterways.

The highest spot is the one where the Dutch Windmill has been built; also the mound of rocks[13] that is in the middle of the body of water closest to this Mill & that forms a cascade, is the point from which the water flows around the garden after having been brought there by the steam pump, which is near the hothouse & by the pump that the Windmill turns.

1. The water is distributed in several ways: after leaving the pool around the rock,[14] it forms a fountain,

---

7. The "people" in question are almost certainly Morel and Claude-Henri Watelet (1718–1786), both of whom lauded the *ferme*.

8. A pastoral novel written by Honoré d'Urfé and published between 1607 and 1627.

9. The author Bernard Le Bouyer de Fontenelle (1657–1757).

10. The painter François Boucher (1703–70).

11. *ils* in the original, referring to "the inhabitants of the country side."

12. *elles* in the original, referring to "French Ladies."

13. The word Carmontelle uses here is "rocher," which in the eighteenth century had a specific meaning when applied to gardens: it suggests a constructed mass or heap of natural rocks, stone, shells, coral, and moss forming an ornamental fountain.

14. Carmontelle's "bassin du rocher" has no satisfactory English equivalent. It is a garden feature with a "rocher" surrounded by a pool of water.

which feeds the stream that runs alongside the flowering water meadow, passes behind the Italian Vineyard, & flows into the Circus, under the wooden bridge.

2. The pool around the rock provides water to the Miller's house, which is a Dairy all in marble; it irrigates its Garden, goes on to feed the fountain that is in the wood of tombs, & joins up, under the wooden bridge, with the river that flows into the Circus pool, after having provided water to the large tomb.

3. The pool around the rock supplies water to the pool on the terrace of the Water Mill, feeds the cascade of the ruined Castle, & falls onto the wheel of the Mill, which turns a pump that is inside this Mill.

4. The pool around the rock feeds the river that goes from the rock to the Mill bridge, via the large aqueduct, descends in a waterfall there & forms several more waterfalls at the level of the Water Mill bridge; from there it flows under the bridge, & by diverging forms the Isle of Sheep before continuing to the Turret at the entrance to the Garden where it seems to disappear beneath an arcade; but it comes back, underground, to the river at the Chinese bridge, from where this river cascades in a double waterfall by the Nymph fountain before flowing into the Circus pool.

6.[15] In the same way the pool around the rock supplies the Duck Pond, which is at the foot of the Minaret hillock.

The water that is above the ruined Castle flows, in part, under the Willow bridge in front of the marble Temple alongside the regular wood, where it forms a canal around the brass-ring carousel,[16] & also the pool in front of the large pavilion; from there it follows the wood & passes in front of the Horse Chestnut chamber,[17] then forms a stream that cascades alongside the wood bordering the rue de Chartres & ends in a fountain on the other side of the ditch, which provides water for the animals' drinking trough.

The majority of the water is brought by gravity to the Circus pool; it flows into the large steam pump, which is nearby in the irregular wood, whence it is sent back to the first rock, whence it flows & divides as described above. The large steam pump also provides water to the bathing female figures in the pool that is near to the Circus & that empties in cascades into the Circus.

It can be seen, by way of this water circuit, that it is easy to irrigate every part of this Garden, & that by firing up the steam pumps, one is master of all this water & can set it in motion whenever one wants.

Through the detailed descriptions of the views, & by following the plan, everything will be even more easily explained.

VIEW

Of the Water Mill, & of the Bridge leading to it, taken from point A, *near the Sundial.*
[PLATE II]

On entering the Garden at Monceau, one finds oneself in a small meadow in which a river runs forming the Isle of Sheep. This meadow is bordered to the right by a wood, & to the left by a court, planted with Lombardy poplars & Horse Chestnuts, which is located in front of the main pavilion. On arriving at point A near the sundial one sees on the left the drawbridge gate of an old ruined Fort, which is attached to an old Tower, also ruined, & to a square, crenelated, brick Building, in which a Water Mill

---

15. The misnumbering is presumably a typesetting error.

16. "Jeu de Bague" in the original French. Whether or not the ring was made of brass is uncertain, but this admittedly anachronistic expression is probably the one that will best evoke Carmontelle's contraption to a contemporary American audience.

17. "Salle des Marronniers" in the original French: an open space surrounded by horse chestnut trees, probably inspired by Jules Hardouin-Mansart's Salle des Marronniers at Versailles.

has been established. Next comes a brick & rubblestone bridge of three arches, & below the main arch the water that comes from further away falls in cascades. Above the bridge, on the horizon, can be seen the hill of Montmartre, which closes the view. Between this hill & the bridge can be seen two pavilions that belong to the wood of Sycamores & Ebony trees, then the Dutch Windmill, the Minaret hillock, the mound of rocks—the source of all the water—& the ruins of the Temple of Mars.

VIEW

Of the entrance to the Garden, taken from point B, *on the Isle of Sheep.* [PLATE III]

After crossing over to the Isle of Sheep, at point B, one can see the entrance to the Garden, which is next to a small Gothic Building that serves as a Chemistry laboratory & has a Turret that terminates in a point. Under the arcade of a crenelated wall the river passes. To the right is the court that is in front of the main pavilion, in the middle of which are several trees, grouped around a perch, which is intended for shooting the popinjay.[18]

The foreign figures who are in these drawings are the persons who serve the Prince in different costumes, when he dines or sups in his pavilion, or strolls in his Garden.

On leaving the Isle of Sheep, to the right, there is a little Fountain set into a rock on top of which there is a ruin. If one passes in front of the Mill bridge & under the drawbridge gateway, & one then climbs atop the Mill, by turning to left & right one can get a good idea of this part of the Garden.

Going down to the right, crossing the Willow bridge & again going right, one will find a sunken path lined with woody nightshade made to provide shelter when there is too much sun. After passing under an aqueduct one will find oneself, having turned left, at point C.

VIEW

Of the ruined Castle, with its Cascade, & of the bridge leading to it, taken from point C,
near the Temple of Mars. [PLATE IV]

Here one sees, above the bridge, the entrance Turret, the top of the principal pavilion, the ruined Gothic Castle, the water that falls on its ruins, the foundations of a large tower, the Willow bridge, & part of the white-marble Temple.

In the foreground the river coming from the large rock is divided by the Isle of Flowers, around which it flows on either side, falling in cascades under the large bridge; & on the right it carries on, flowing at the foot of the Gothic Castle, then passing under the Willow bridge, & continuing its course in front of the marble Temple, before running alongside the regular wood.

Passing in front of the Temple of Mars & turning right along the path, one arrives at point D.

VIEW

Of the Farm, taken from point D, *near the Cabaret.* [PLATE V]

One sees the Buildings on the left, of which one can make out only the top of the Gardener's dwelling & the cow sheds; & those in the foreground are the goat & sheep sheds. At the rear is the

---

18. "Tir à l'oiseau" in the original French, a game of target practice that involved shooting at a colored wooden bird. In the British Isles the colored bird was traditionally referred to as a papingo or popinjay.

(18)

dovecote behind which are the hothouses, the steam pump, & the fig orchard; & in front, the wood containing the little rock fountain that was mentioned in the description of Plate III.

Turning around, one can take the path to the right, running alongside the Monceau road, & one will arrive near point E.

VIEW

Of the ruins of the Temple of Mars, taken from point E, *near the Monceau road.* [PLATE VI]

One sees the ruins of the Temple of Mars, whose columns are of the Corinthian order. This Temple appears to have been square & to have had a peristyle, of which one sees two parts. In the middle there was a statue of Mars that was too disfigured; it has been replaced by one of Perseus, which is antique. The Temple has none the less kept the name of Temple of Mars.

In the distance, on the left, can be seen a part of the same things as in Plate IV, & on the other side is the Isle of Rocks, & in the distance can be seen the Tartar[19] Tent & the two pavilions backing onto the wood of Sycamores & Ebony trees.

CHAPTER II.

VIEW

Of the Isle of Rocks & the Dutch Windmill, taken from point F. [PLATE VII]

FROM point F one sees, on the left, the little rustic bridge by which one enters the little Isle of Rocks, before passing into the meadow containing the wood of Sycamores & Ebony trees, as well as the three pink, yellow, & blue Gardens, with the two French Pavilions. On the right, atop the hillock, is the Dutch Windmill that turns a pump, which partly supplies the cascade that is next to the mill, & behind the mill is the Miller's house, which forms a Dairy, decorated inside in marble, & outside very rustically. Above the rock, from the spring, one can make out the Minaret in the distance.

VIEW

Of the Minaret hillock, taken from point G. [PLATE VIII]

After coming out of the Dairy, climbing back up the Windmill hillock & going round the pool of the rocks, one will arrive at point G, whence one will see the Minaret hillock, planted with vines, among the rocks of which it is constituted, & on the right part of the chestnut grove. Turning left, one will pass in front of a small fountain, which feeds the little stream that runs around the flowering water meadow. Alongside this stream there is a pussy willow path, which exactly follows its contours; following this path, one arrives at the icehouse bridge, & on turning right one will find a Lair formed by some of the rocks that support the Minaret hillock. If one climbs up to the Minaret &

---

19. The traditional spelling "Tartar" has been used in this translation, rather than the more contemporary "Tatar."

(19)

scans the horizon with one's eyes, one will see, beginning at Montmartre, to the right the heights of
Belleville & all the monuments of Paris, finishing at the Observatory. Then Vanves, Issy, Meudon,
Bellevue, Sêve,[20] Saint-Cloud, the Mont-Valérien, the heights of Marly, Saint-Germain, the heights
of Sanois,[21] those of Saint-Prix, Montmorency, & below the Château de Montmorency, Ecouen, &
on the river bank, Epinay, then Saint-Denis, & then Montmartre again.

VIEW

Of the Italian Vineyard, taken from point H. [PLATE IX]

Coming down from the Minaret, on the side one went up, crossing the icehouse bridge, & following
the path that runs along the flowering water meadow, one arrives at point H, from which one can see the
Italian Vineyard.

These vines climb up posts, the vine stocks being planted in staggered rows & the posts holding
up a trellis across which the vines spread at a height of seven feet so that their grapes can be picked.
This trellis maintains an opening of three square feet between the trunks, allowing the sun to reach
the bottom of the vine. In the middle of this vineyard there is an antique statue of Bacchus.

Going back up the same path, crossing the icehouse bridge & turning left, one enters a wood
where there are three irregular routes, two of which lead to a resting spot where there is an
antique statue of Mercury. Following the path, one arrives in front of two ruined monuments,
inside one of which is the large steam pump & a little chamber decorated in the Chinese style.
Continuing along this route, one comes to a little court; there is a pool surrounded by three steps
where one can see the figure of a woman in white marble, who is bathing & a bronze Negress,
who is pouring water onto her body. These two figures are by Mr. Houdon.[22] Turning around,
one goes down a grass path lined with turf benches whose backs are topped by lilacs; retracing
one's steps, through the ruins one sees the Naumachia;[23] in the middle there is a Granite obelisk
on which there are all the Egyptian characters found on the one in Heliopolis. Turning left, under
a peristyle of Corinthian columns & following the wood, one will cross a bridge, from where,
following the path, one will arrive at point I.

VIEW

Of the wooden Bridge, taken from point I. [PLATE X]

One can see the wooden bridge set on rocks that elevate the terrain & form a carriageway. On the
left, near the bridge, is a milestone; above the carriageway one can see the peristyle that forms the
Circus or Naumachia; behind the columns are three tiers of flowers. Taking the same path up to the
carriageway, turning left, & crossing the wooden bridge, near the milestone, one will reach point K.

VIEW

Of the Circus or Naumachia, taken from point K. [PLATE XI]

One sees the whole Circus, in the middle of which is the obelisk. To the left is the river that fills

---

20. Today known as Sèvres.
21. The modern spelling is Sannois.
22. Jean-Antoine Houdon (1741–1828). The marble figure of the bather from this fountain is in the Metropolitan Museum of Art.
23. In the ancient Roman world, both the staging of naval battles and the pool or building in which this took place.

(20)

the pool. To the right, above the regular wood, can be seen the top of the ruins containing the large steam engine & a bell tower expressly designed to serve as a viewpoint over the wood. To the left of the wooden bridge is a milestone.

Turning around & following the carriageway, one finds, to the left, a path that crosses the wood of tombs. Before crossing a little stream, on the left, one arrives at point L.

VIEW

Of the wood of Tombs, taken from point L. [PLATE XII]

This wood is made up of Lombardy Poplars, Sycamores, Cypresses, Plane Trees, & Chinese Thujas. The main pyramidal tomb is Egyptian. The interior is adorned with eight granite columns that are one-third buried & have capitals that are adorned with Egyptian heads holding up an entablature of white marble, granite, & bronze. The vault is adorned with bronze roses. To the right & the left are two tombs of antique black marble; facing you, opposite the doorway, is a niche containing a small basin of antique green marble; in this basin there is the figure of woman sitting on her heels who is squeezing her breasts, from which comes the water that falls into the basin. This figure is Egyptian, in the most beautiful black, & her headdress consists of a silver headband & silver ribbons. In the corners are four niches containing bronze incense burners. The entrance is closed by a gate & the door jambs are two Egyptian caryatids carrying a piece of antique green marble that serves as a lintel.

To the left of this tomb is a bronze urn on a marble pedestal surrounded by four steps. To the right a ruined fountain can be seen, near which is a mulberry. This wood could be the setting for Pyramus & Thisbe, there being another tomb here on which is a ruined obelisk; & on the other side, the tomb of a young girl, in black jasper, which is also damaged. A small stream runs through this wood before emptying into the large pool of the Circus.

CHAPTER III.

VIEW

Of the two French Pavilions, taken from point M. [PLATE XIII]

AFTER seeing the wood of tombs, if one goes back along the path that passes in front of the blue Garden, & that connects up to the main path, which passes in front of the pink Garden, one will easily find point M.

The two French pavilions are painted to resemble marble, with pilasters whose granular moldings are gilded, as are the interior adornments & those on the roof. These pavilions stand in front of the wood of Sycamores & Ebony trees: in the middle of this wood is the Plane Tree grove, so called because at its center there is a Plane Tree surrounded by tiers of flowers. Three marble vases decorate this grove; near the wood, on the tomb side, between the blue Garden & the yellow

Garden, there is an antique statue of Meleager. On the windmill side, between the yellow Garden
& the pink Garden, there is an antique statue of Hymen, & between the pink Garden & the blue
Garden, there is a statue of Friendship by Mr. Pigale;[24] it is this statue that can be seen in this view.
Turning back, one can pass through the Tartar tent & then go down to the river's edge, cross a little
swing bridge, follow the sunken path a few steps, & when one comes out, one will find oneself in a
meadow at point N.

VIEW

Of the Tartar Tent, taken from point N. [PLATE XIV]

This tent is round & lets in light from the top, in the manner of the Tartars who lay fires in the
middle of their tents. It is surrounded by Lombardy Poplars, Sycamores, & Elders. On the left can
be seen the Nymph's fountain, & in the distance the top of the two French pavilions.

The hedge in the foreground, & which occupies the entire width of the picture, is woody nightshade;
it covers part of the sunken path along which one can walk in shade from the Chinese bridge to the pool
of the Circus & in this way reach the irregular wood under shelter. If one enters this path & follows it,
on the southern side, one will end up at the Chinese bridge; crossing it, & keeping the irregular wood to
the right, one finds oneself in front of the white-marble Temple, at point O.

VIEW

Of the white-marble Temple, taken from point O. [PLATE XV]

This Temple, although it is whole, is surrounded by trees, which makes the effect more piquant
through the greenery. The edge of the wood is encircled with flowering shrubs, which form a border to a little stream that must be crossed to enter the Temple. A small rotunda, it is entirely open
to the light at the top, & between the twelve Corinthian columns of which it is formed there are
marble benches. Placed on the altar, which is in the middle, is an antique figure representing one of
Achilles's female companions when he was at the court of king Lycomedes. This figure is holding
one of the gifts sent by Ulysses from among which Achilles chose a helmet & a sword, while his
female companions preferred adornments with which to bedeck themselves

Retracing one's steps, crossing the Chinese bridge & turning left, one arrives in a meadow
where there are groups of Sycamores & Plane trees. This meadow, on the left, is delimited by a
stream that follows the regular wood, in which there are only groves of trees, in the taste of those
found in ordinary Gardens. By the principal pavilion between two galleries can be seen a fairly
large aviary, but of a type whose effect would be difficult to show in a view. Strolling in the meadow, one finds the Whitebeam grove on the right; next there is a very small grove of trees & shrubs,
in which a little porcelain Amour has been placed; nearby is point P, from which the carousel & the
principal pavilion can be seen.

---

24. Jean-Baptiste Pigalle (1714–85). The statue in question was probably his *Madame de Pompadour en Amitié* (1753), which is now in the
Louvre.

VIEW

Of the principal Pavilion & of the Carousel, taken from point P. [PLATE XVI]

This pavilion was made to be extraordinary. When a new form was given to the Garden, changes had to be made to it (1). To make it more agreeable, two pediments were destroyed which, when seen from the side produced an awkward effect; it would have been desirable to add a frieze to the entablature. The roof was painted to resemble stone, with bronze garlands, & pilasters were added all around the building, rusticated with blocks of yellow marble from Siena & plaques of Languedoc marble. The capitals, the bases, the adornments, & the moldings are in antique bronze. To extend the pavilion, four galleries were added, each seven bays wide & surmounted by an antique balustrade & adorned with plaques in violet marble.[25] Before the building is a pool, which runs in a circle around a brass-ring carousel, enclosing it in an Island. The carousel is a Chinese parasol held up by three Chinese figures,[26] which also hold a horizontal bar that is used by those who turn the carousel, & who need do nothing other than walk on the wooden boards beneath their feet. From the edge of the boards, four iron branches extend, of which two support dragons that can be mounted like a horse. At the end of the other two branches there are reclining Chinese costumed mannequins; with one arm they support a cushion on which one may sit, & each holds in his hand a parasol trimmed with little bells; with the other hand they hold a cushion for the feet. Ladies sit on these two branches.

The edge of the large parasol is trimmed with ostrich eggs & bells. The four lanterns that can be seen contain rings at the end of the bobbles beneath the lanterns that are only visible to those riding the carousel.

The Chinese bridges over which one passes to enter the Island carry on round to make balustrades in front of the benches, which are in the form of Ottomans. These benches are in stone & imitate Persian tiles. Above these Ottomans are draperies held up by poles. These draperies are striped with violet, saffron, & white. This is where the company sits to watch the carousel. To the right & the left of these Ottomans are vases in red bronze, of which the garlands & all the adornments are gilded. Leaving the Ottoman, which can be seen on the left, following the pool, & coming round to the side of the main pavilion, one will arrive at point Q, at the entrance to the two groves.

VIEW

Of the Turkish Tents, taken from point Q. [PLATE XVII]

The part of the gallery seen on the left was built to link with the billiard room & give rise to the four galleries; but the billiard room at the end produced an awkward effect, & it was decided to decorate it as a Turkish tent & to add the round tent that is in front of it. These tents are striped, the first in red & white, the second in blue & white; these stripes are damasked & the adornments are gilded. The tents are in a grove bordered with flowers that stretches as far as one of the Ottomans to be seen on one side, as well as to the bridge that accesses the Carousel Island.

Passing in front of this Ottoman, one enters the rest of the meadow through which the carousel was gained, & one arrives at point R.

1. It is unknown who provided the design.

---

25. "Brèche" in the original French, a term used to designate certain types of marble.

26. *Pagodes*, in the original French, are porcelain figurines often with bobbing heads, best understood here as mannequin-like figures dressed in Chinese costumes.

(23)

Of the Horse Chestnut Chamber, taken from point R. [PLATE XVIII]

This chamber backs onto the wood that runs all along the rue de Chartres; it is adorned on the outside with an antique balustrade on which are hung blue & gold carpets. One climbs three steps to enter the chamber. At the back is a niche, set between two columns holding up an entablature. The architecture is in the proportions of the Doric order, although without triglyphs, & the columns are rusticated. In the middle of the chamber is a basket of flowers, & on a pedestal in the niche can be seen a statue in white marble, a copy by Bouchardon of the antique Faun from the Villa Borghese.[27] Around the chamber there is a bench with a turf seat back, & the seat back is crowned with flowers. The view from this chamber dominates the surrounding countryside; without realizing it, one has completed the visit to the Garden.

END.

Read & approved, 28 March 1779. DE SAUVIGNY.
Whereas, Approval has been granted, license to print, LENOIR

From the Printing House of L. JORRY, rue de la Huchette.

---

27. The French text reads "Vigne Borghese," an alternative name for the Villa Borghese used occasionally in the eighteenth century. The faun at Monceau, however, was not a copy of the one in the Borghese collection, but rather of the Barberini Faun. The Monceau copy is now in the Louvre; the original is in the Glyptothek, Munich.

PLATES

PLATE I

The plan of the Garden at Monceau.

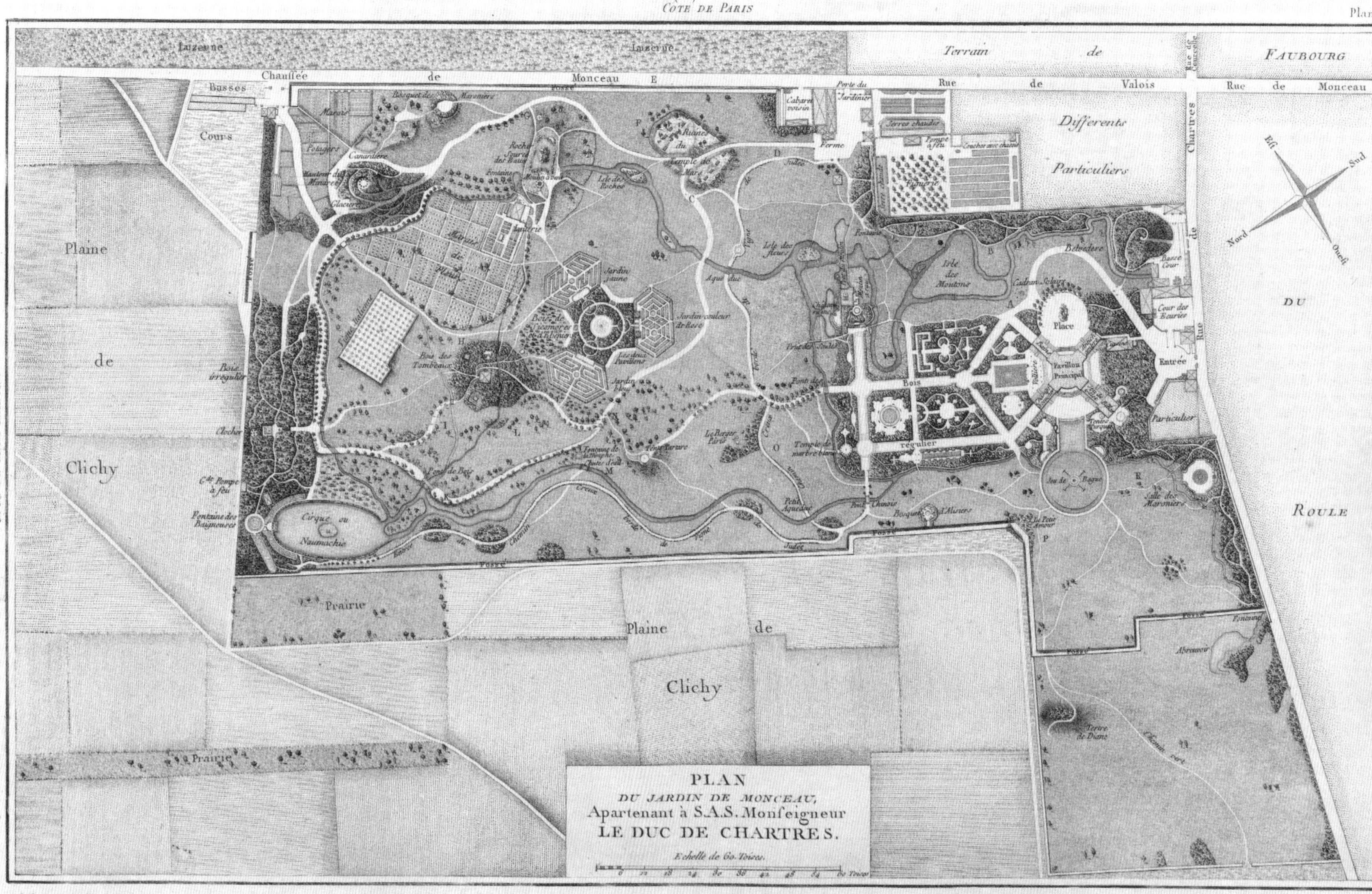

PLAN
DU JARDIN DE MONCEAU,
Apartenant à S.A.S. Monseigneur
LE DUC DE CHARTRES.

Echelle de 60 Toises.

PLATE II

Of the Water Mill, & of the Bridge leading to it, taken from point A, *near the Sundial.*

VÜE.

Du Moulin a eau, et du Pont qui y Conduit,

Prise du Point A, proche le Cadran Solaire.

L. C. de Carmontelle Inv. del.

J. Couché Sculp.

PLATE III

Of the entrance to the Garden, taken from point B, *on the Isle of Sheep.*

L. C. de Carmontelle Inv. del.

J. le Roy Sculp.

VÜE
De l'Entrée du Jardin et du Principal Pavillon.
Prise du point B. dans l'Isle des Moutons.

PLATE IV

Of the ruined Castle, with its Cascade, & of the bridge leading to it, taken from point C, near the Temple of Mars.

Planche . IV.

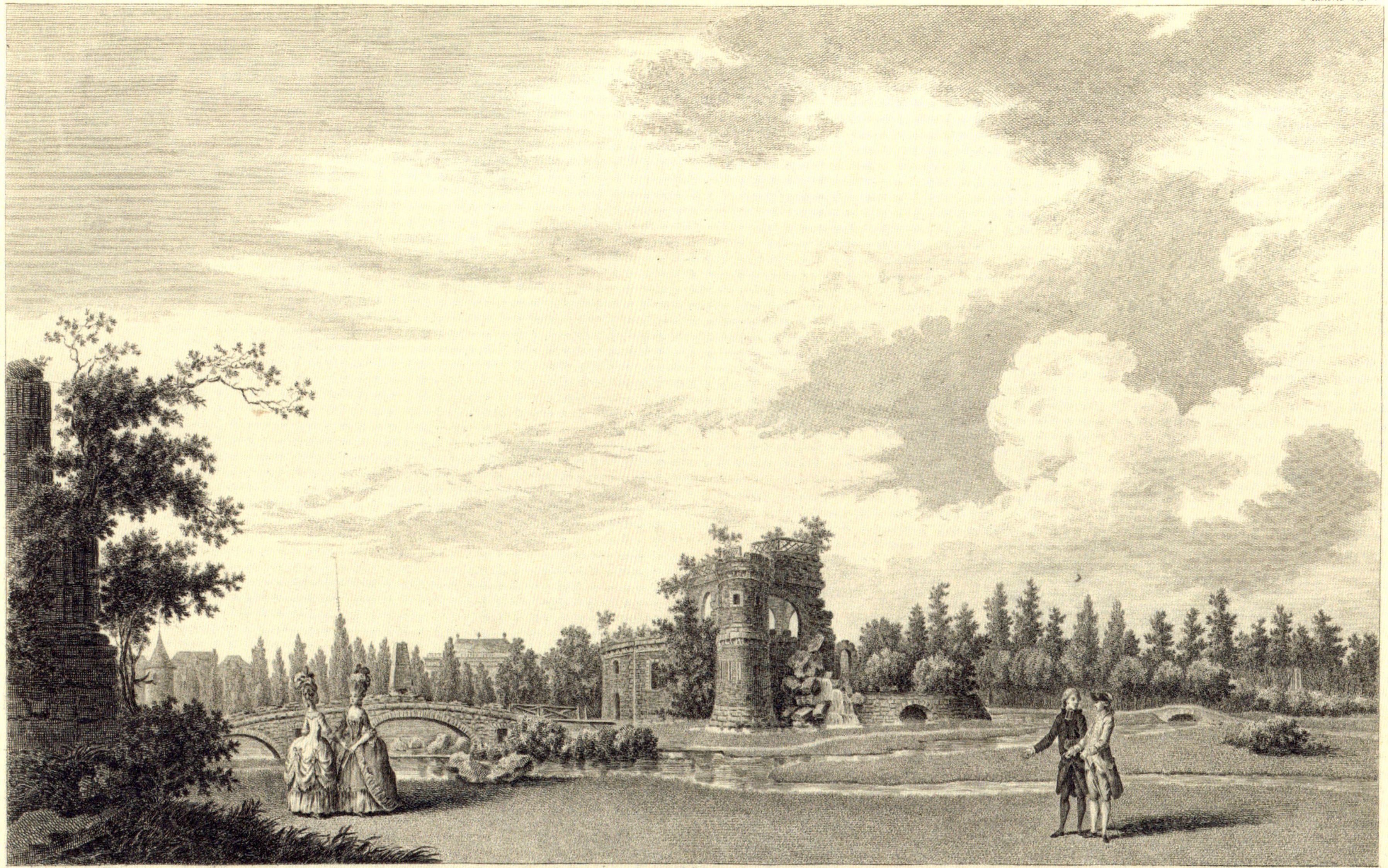

L. C. de Carmontelle inv. del.

T. Doni Sculp.

VÜE

du Château ruiné, avec sa Cascade et du Pont qui y conduit,

Prise du point C. proche le Temple de Mars.

PLATE V

Of the Farm, taken from point D, *near the Cabaret.*

L. C. de Carmontelle, inv. del.

O. Michel, Sculp.

VÜE
de la Ferme,
Prise du point D. auprès du Cabaret.

PLATE VI

Of the ruins of the Temple of Mars, taken from point E, *near the Monceau road.*

L.C. de Carmontelle inv. del.

E. pine Sculp.

VÜE
Des Ruines du Temple de Mars.
Prise du point E. près du Chemin de Monceau.

PLATE VII

Of the Isle of Rocks & the Dutch Windmill, taken from point F.

L. C. De Carmontelle inv. del.

Michault Sculp.

VÜE
De l'Isle des Roches et du Moulin hollandois,
Prise du Point F.

PLATE VIII

Of the Minaret hillock, taken from point G.

VÜE
De la hauteur du Minarêt,
Prise du Point G.

PLATE IX

Of the Italian Vineyard, taken from point H.

J. C. De Carmontelle Inv. del.

Michault Sculp.

VÜE
De la Vigne Italienne,
Prise du Point H.

PLATE X

Of the wooden Bridge, taken from point I.

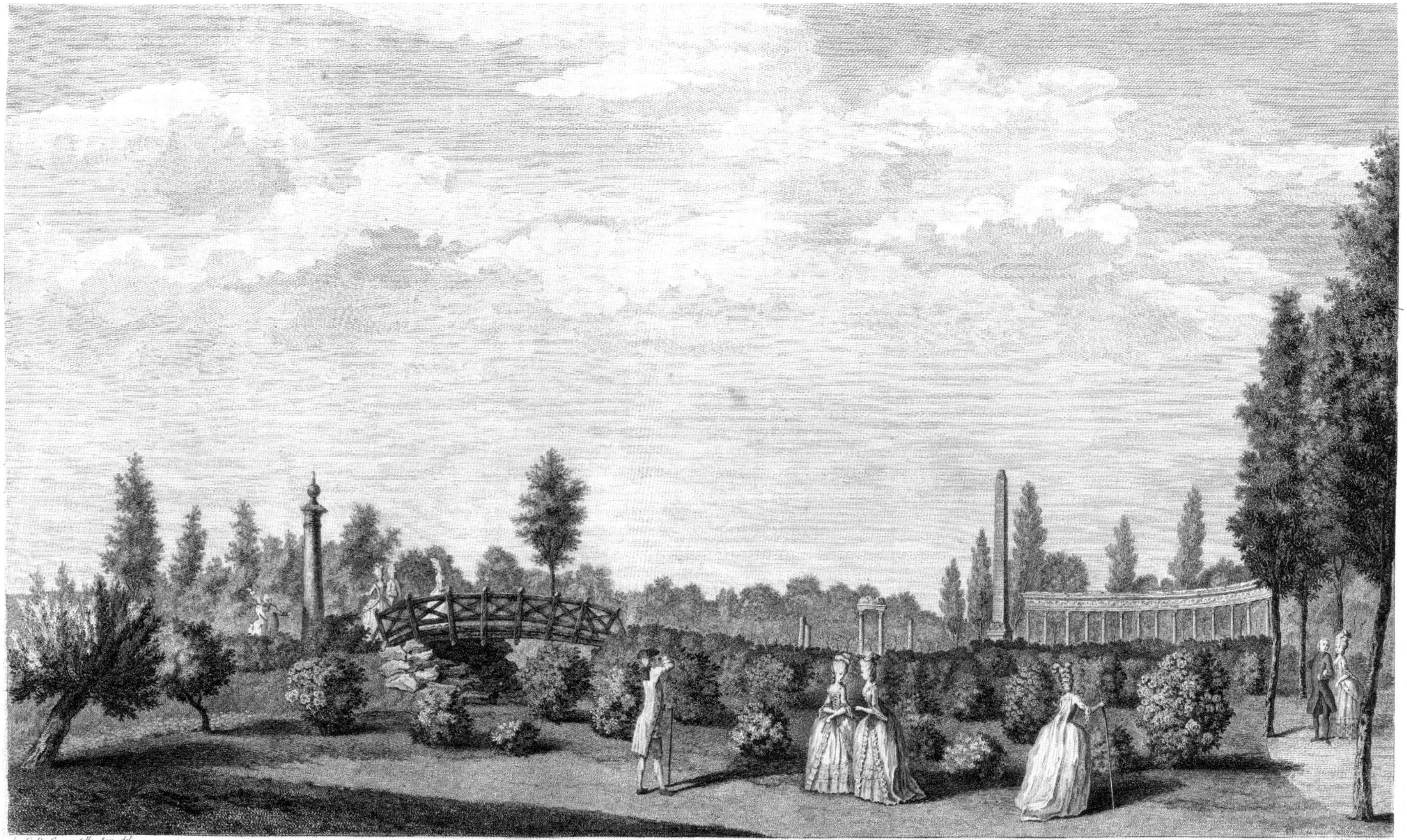

L. C. De Carmontelle Inv. del.

Colibeon Sculp.

VÜE
Du Pont de Bois,
Prise du Point X.

PLATE XI

Of the Circus or Naumachia, taken from point K.

L. C. De Carmontelle Inv. del.

Lépine Sculp.

VÜE
Du Cirque ou de la Naumachie,
Prise du Point K.

PLATE XII

Of the wood of Tombs, taken from point L.

L. C. De Carmontelle Inv. del.

L. Lerouge Sculp.

VÜE
Du Bois des Tombeaux,
Prise du Point L.

PLATE XIII

❧

Of the two French Pavilions, taken from point M.

L. C. De Carmontelle inv. del.

Mishault Sculp.

VÜE
des deux Pavillons François,
Prise du Point M.

PLATE XIV

Of the Tartar Tent, taken from point N.

L. C. De Carmontelle. Inv. del.

Legrand Sculp.

VÜE
De la Tente Tartare,
Prise du Point. N.

PLATE XV

Of the white-marble Temple, taken from point O.

L. C. de Carmontelle inv. del.

J. le Roy Sculp.

VÜE
Du Temple de Marbre blanc,
Prise du point O.

PLATE XVI

Of the principal Pavilion & of the Carousel, taken from point P.

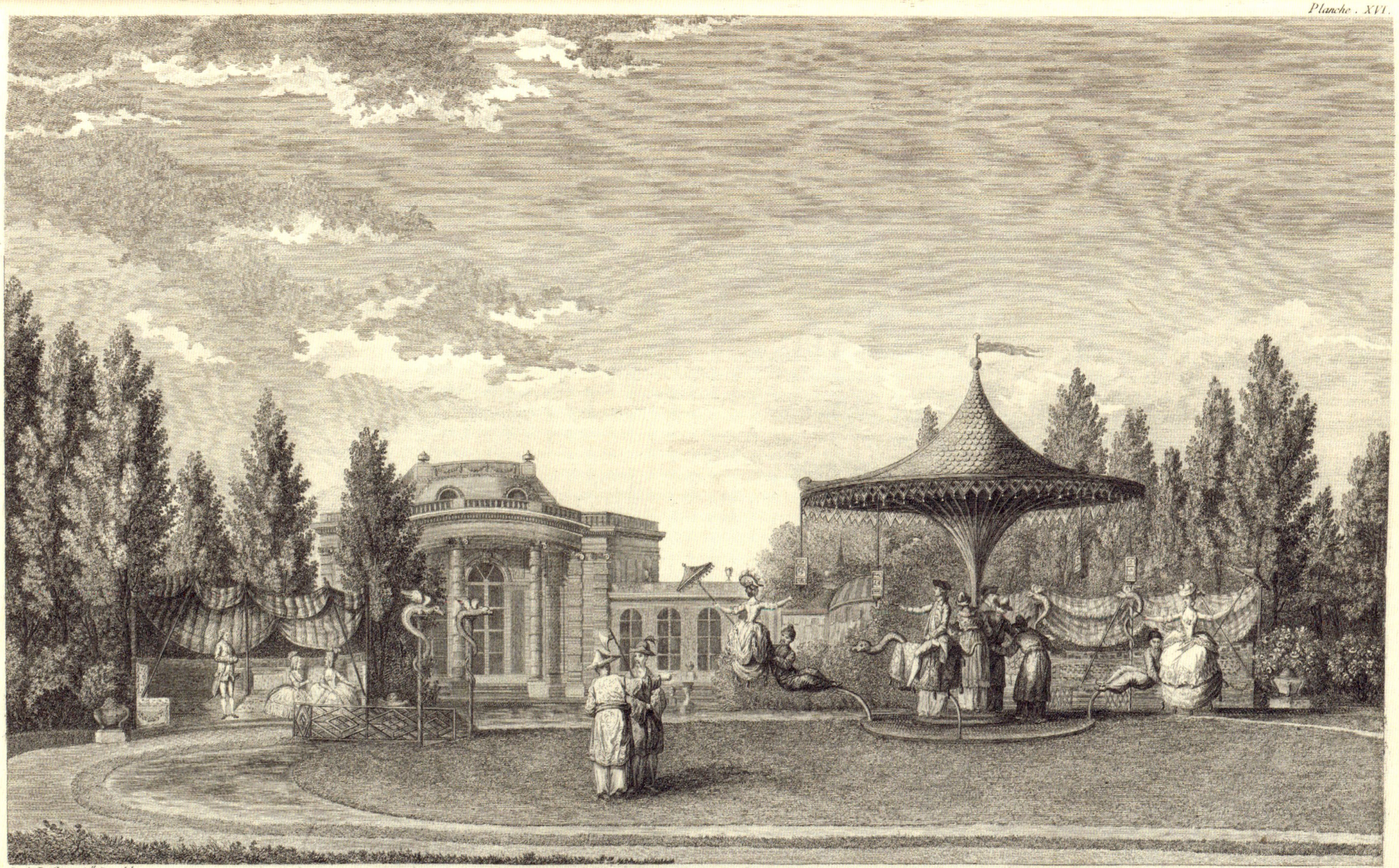

L. C. De Carmontelle Inv. del .

Michault Sculp .

VÜE
Du principal Pavillon et du jeu de Bague,
Prise du Point P.

PLATE XVII

Of the Turkish Tents, taken from point Q.

L. C. De Carmontelle Inv. Sculp.

Mathault Sculp.

VÜE
Des Tentes Turques,
Prise du Point. Q.

PLATE XVIII

Of the Horse Chestnut Chamber, taken from point R.

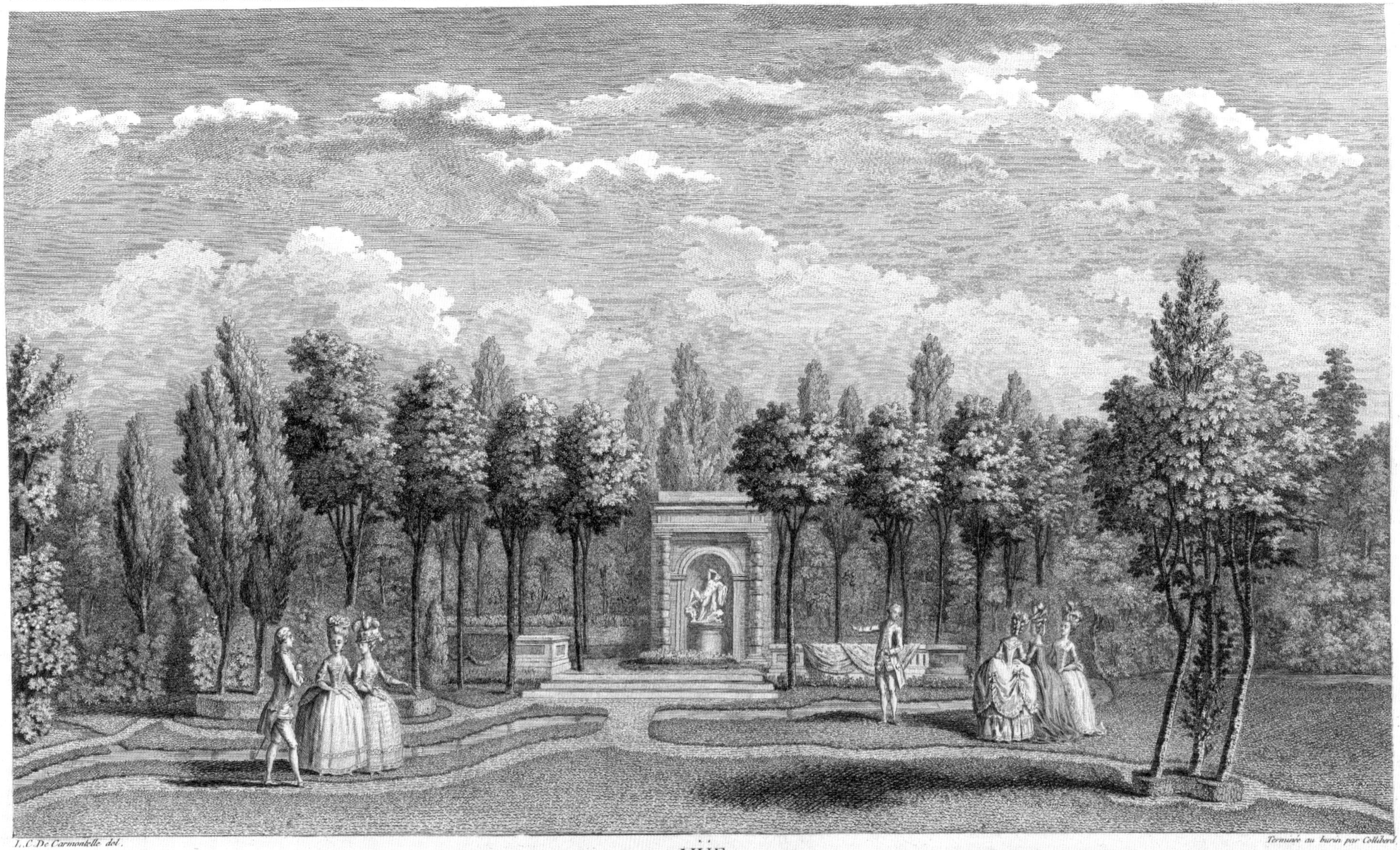

L.C. De Carmontelle del.

Terminée au burin par Collibert

VÜE
De la Salle des Marroniers,
Prise du Point. R.

CARMONTELLE AND HIS WORLD

Laurence Chatel de Brancion

Agarden is a setting for the wanderings of "our soul, which desires only sweet, lively, & gay impressions"; an intellectual creation that abounds with "fortuitous negligences, which belong only to taste" (*Garden at Monceau*, 15): thus wrote Louis Carrogis, known as Carmontelle. His garden at Monceau—an extraordinary folly created for Louis-Philippe-Joseph d'Orléans—was his own great invention. Alas, we shall never see Carmontelle's Monceau: ten years after its creation it was utterly changed. This essay and the essays that follow in this volume examine Carmontelle's garden, its creator, and his world.

I have spent many hours with Carmontelle: exploring the hundreds of notable figures and characters he painted and drew, and those he placed in plays, novels, and even in "films" that anticipated today's cinema. In them the ancien régime breathes "in a circle of illusions and in a sort of joy"—as the comte de Ségur put it—or expresses a "plaisir de vivre" as whispered the prince de Talleyrand.[1]

My windows look over the greenery of the Parc Monceau. Today there is little left of the extraordinary encounter between a lively imagination and a prince's fortune. Nonetheless, if I walk in the park, an apparition around a bend—a pyramid or a tumbledown colonnade by a pool of water—raises a question: "What was Monceau?"

CARMONTELLE

Louis Carrogis was born in 1717 in the faubourg Saint-Sulpice on the outskirts of Paris, the son of a cobbler who worked in leather from his native Pyrenees. Carrogis was probably taught by the Jesuits at the Collège Louis le Grand. Besides receiving a rigorous classical education, he would have been schooled in the art of theater, which was considered indispensable for accustoming the young to public life. Later he was appreciated for his poise and cultivation, which allowed him to converse brilliantly in society salons.

As a young man Carrogis became an army topographer, adopting the name Carmontelle (a sobriquet whose origins have not come down to us) at some point over the course of his military career. Proudly he added the title of engineer to his name when he signed the baptism register of his nephew and godson at

NOTE. *Page numbers following quotations by Carmontelle throughout this book refer to the translations in the current volume.*

FIG. 1 The duc d'Orléans and his son, the duc de Chartres, in 1759, by Carmontelle.

FIG. 2 An evening at the Palais Royal, ca. 1773–75, by Carmontelle.

at Saint-Sulpice. He learned to paint in the army, claiming Francesco Casanova, a painter of battles and the brother of the notorious seducer, as his master.

During Carmontelle's military career he drew up detailed maps. He reconnoitered terrain, the course of rivers, and sources of water; he assessed inclines and analyzed fortifications. In so doing he acquired a smattering of ideas about architecture. All in all, it was an excellent training for a future garden designer. "I used to be an engineer and no one could draw more skillfully," he wrote. "You should have seen me at a siege, I always went unprotected, nothing stopped me; once I was off, I would have crossed all defense barriers."[2] During the inevitable periods of inaction—those empty days that most soldiers filled with gambling and wine—Carmontelle took up his pencils and drew the officers in his regiment. His artistic process rapidly became an entertainment, with military men gathering to watch as he sketched one of them, capturing the details that brought out their resemblance. Afterwards they would admire and critique his work, comparing the portraits affixed to the gun carriages to the subjects. Soon the officers of neighboring regiments heard of his talents and sought him out. He also entertained his comrades by writing and staging short plays in which the underemployed soldiers became actors. Thus Carmontelle became known to, and appreciated by, the marquis d'Armentières, who hired him to teach the art of war to his sons, after which the duc de Luynes entrusted his heir to him.[3] Finally the duc d'Orléans called upon his services.[4]

Louis-Philippe de Bourbon, duc d'Orléans, first prince of the blood (1725–85), wanted his only son, Louis-Philippe-Joseph, duc de Chartres (1747–93), to possess the knowledge needed by a prince whose destiny was to command—a role that, at the time, principally involved laying siege to a town (fig. 1). Carmontelle was hired "for the use of which it was believed he could be to the prince in mathematics, fortifications, and drawing. He could, in the empty moments of the day, place under his gaze what the masters taught him without its having the disagreeableness which the whiff of lessons and study nearly always has for the young"[5]—a remark that insinuates a certain nonchalance in the duc de Chartres's character. The prince was ten years old; his mother, born a princess of Conti, was about to die at the age of 32; and "at this same moment, M. de Carmontelle was assigned to his person as a tutor."[6]

His father, now a widower, carried out military obligations in times of war; was a keen and proficient hunter; and in peacetime amused himself in the company of loose women. While Chartres's sister, Bathilde, was sent to a convent to be educated, Louis-Philippe was entrusted to a governor, the elegant and haughty comte de Pons, who brought him up to be cognizant of his role and convinced of his superiority. He acquired perfect deportment and a sharp understanding of protocol and the workings of the court, and became accustomed to living constantly in the public eye. But while Pons perfectly initiated him in worldly comportment, he neglected his intellectual and moral education completely. Talleyrand would later say, "No one paid any more attention to his character than to his studies. Since he had an elegant waistline, they tried to get him to excel at physical exercises. Few youths rode as well and as gracefully as him. He was adept with weapons; at balls, everyone looked at him."[7]

Likenesses

So it was that Carmontelle, in the winter of 1759, moved into the heart of Paris, to the Palais-Royal, the official residence of the princes of Orléans. At the end of Louis XV's reign, and even more so during Louis XVI's, dignitaries, unless they had particular obligations, no longer went to Versailles except on Sundays (to attend the king's mass and the reception that followed it). The Palais-Royal consequently strengthened its allure, as much for the courtiers surrounding the first prince of the blood as for the general public, the gardens being open to everyone all day long. The palace contained an exceptional art collection, amassed by Philippe d'Orléans, Louis XIV's brother, and augmented by his son, the Regent of the kingdom during Louis XV's minority. In his *Description historique de la Ville de Paris*, Piganiol de La Force notes that "the great knowledge that the duc d'Orléans, Regent, had of painting caused him to seek out and buy, wherever he could find them, the most excellent pictures by the great painters, so that the repository he left is the richest and most curious in the world, not even excepting the king's."[8] The Regent reorganized the apartments at the Palais-Royal to serve as a setting for his collection and opened it to visitors; young painters rushed to get permission to work there, and all enlightened enthusiasts visiting Paris included the Palais-Royal on their list of sites to visit. Arriving from America, William Lucas reckoned one would need to spend two

FIG. 3 Courtiers in *habit de campagne* at the Orléans estate of Saint-Cloud, by Carmontelle.

hours a day there for six months, while William Cole acknowledged that "this gallery alone is worth a long voyage"![9] Every day, Carmontelle crossed these rooms, admiring works by Raphael, Rembrandt, and Titian and contemplating Poussin's *Seven Sacraments* as he went. It was an entirely unexpected opportunity to educate his eye. He also had access to the library, and here he took his pupil, to whom he taught not only drawing but also engraving.

As a meeting place in the capital the Palais-Royal was visited by every celebrity Europe had to offer. The duc d'Orléans had liked watching Carmontelle sketching his officers in the garrison, and asked him to do the same at the palace. "Carmontel [*sic*], the tutor of M. le duc d'Orléans [then the duc de Chartres], came into the salon after dinner," remembered madame de Genlis. The prince was very fond of him and "it was agreeable to talk with him . . . He was one of the rare few who, not being allowed to dine with the princes, were called in to eat ice cream."[10]

It wasn't only to savor the sorbets then so fashionable with the wealthy that he was invited—it was also to amuse the guests as he had done before with the idle officers. He was supposed to take his pencils and brushes and draw one or another of the people gathered. Was it the duc d'Orléans who suggested he enhance his drawings with watercolor and sometimes a touch of gouache? Or did Carmontelle himself decide to go for color? Either way, from this point on his portraits, which until then had been exclusively drawn with three types of pencil, were completed in watercolor.

On such occasions the assembled company would peer over his shoulder to compare his likeness to the subject, commenting on the portrait in progress. Carmontelle had to put up with this chatter. It sometimes included valid opinions but too often consisted of exasperating blather. Such situations were the theme of *Le Portrait*, one of his *proverbes dramatiques* (little dramas that each riffed on a proverb). In it a portrait painter complained about his clients to a colleague, a history painter:

> "And the likeness that they're never happy with?"
> "Ah! They'll just have to accept it!"
> "That's easy for you to say—it's obvious you don't paint portraits . . ."
> "To the devil with it I say!"
> "I've often been tempted to say the same, but one has to earn a living."
>
> ["Et la ressemblance dont on n'est jamais content?"
> "Ah! Qu'ils s'accommodent!"
> "Cela vous est bien aisé à dire; on voit bien que vous ne peignez pas le Portrait . . ."
> "J'enverrais le métier à tous les diables!"
> "J'en ai été tenté bien des fois; mais il faut vivre."][11]

One had to earn a living and so put up with clients' demands. Carmontelle faithfully captured an evening at the Palais-Royal in a large watercolor in which he adroitly rendered the ambience through spatial relations (fig. 2). The patriarch, the duc d'Orléans, is sitting to the right, a book open in his hands, and is turning towards his daughter-in-law, the duchesse de Chartres, who is seated next to him. Leaning on the duchess's seatback is the comtesse de Genlis, the mistress of the duc de Chartres (a liaison the duchess was unaware of at the time—but Carmontelle wasn't). Leaning on the duc d'Orléans's seatback is *his* mistress, the marquise de Montesson. There is a noticeable likeness between the two women, who were aunt and niece, each being posed in a way suggesting ease and mastery as they frame the duc d'Orléans and duchesse de Chartres. To the left, the duc de Chartres is standing, holding a cane and apparently lecturing Mme de Montesson, while at his feet Mme de Blot, seated on the floor, is looking at the patriarch as is the marquise de Clermont, who is seated at the desk. Behind Chartres an oh-so-fashionable parrot is perched on the hand of the comtesse de Polignac, the queen's favorite. The two men are wearing the *habit de Saint-Cloud* (Saint-Cloud being the Orléans' suburban estate), the father seated—as is the dog in the corner of the image—the son standing like the second dog near his own feet, which is up on its hind legs begging![12] And if we turn our attention once more to the duchesse de Chartres and Mme de Genlis behind her, we notice they are both looking towards the duc de Chartres. Their expressions are innocuous, but their positioning and the direction of their gazes are not, no more so than Mme de Montesson's expression as she listens to Chartres. Her arm, resting nonchalantly on the seatback of the duc d'Orléans, indicates her ascendancy.

Reaching beyond resemblance

Carmontelle had a gift for capturing his sitters' characters. Rather than seeking mere resemblance, he attempted to define identity through attitude, clothing, and details. The reaction of the officers he drew in the army had testified to his talent as a portraitist; now he received renewed confirmation of his gifts daily. Did not Diderot himself, whose writings on art were becoming highly influential, praise him in his account of the 1761 Paris Salon? With respect to Doyen's portrait of a young Indian from Tangiaor, Diderot remarked that he "preferred the profile done of her by M. de Carmontelle, which is truer and more agreeable."[13]

The aesthete Horace Walpole, who had grown up with the painting collection assembled by his father, the British prime minister Sir Robert Walpole, wrote to Mme du Deffand: "Here I am the happiest of men—I have just received the picture. I tore off the wrappings in which it was barricaded, and at last I found you—yes, you yourself. I knew . . . Carmontelle would paint you better than Raphael was ever able to capture a likeness; and it turns out I was spot on. You are here with me in person; I'm talking to you; all that is missing is your impatience to reply. The tulip, your barrel [as she called her ample armchair], your furnishings—everything is there, and in the greatest truthfulness." To which Walpole added this significant remark: "Never has an idea been so well rendered."[14]

One example of Carmontelle's work that has become famous through its use in advertising is a large watercolor representing six of the duc d'Orléans' courtiers wearing the red *habit* "that he had adopted in his country estates" (fig. 3). Carmontelle "had rendered them strikingly even though seen from behind"—the men were recognizable to everyone![15]

FIG. 4 The duc d'Orléans, after Carmontelle.

Carmontelle depicted the duc d'Orléans on horseback in hunting costume, a horn slung round his chest and a landscape in the background, symbolizing his principal occupations of landowner and hunter (fig. 4). Humor is evident in the portraits of Mme d'Epinay's family. All its members are wearing striped clothing to protest the striking off for nonconformism of the barrister Linguet (*rayé* meaning both "striped" and "struck off").There is also comedy in his depiction of the marquis de Flavigny kneeling on a *voyeuse* (a low chair designed for peering indiscreetly over the shoulders of cardplayers). He is placed in front of the skirts and worktable of a lady whose face we do not see—he had seduced so

(67)

many one wouldn't know which to portray! But Carmontelle didn't judge; he depicted. He might portray the comte d'Adhémar with a dilapidated cottage in the background—a nod to the count's humble origins—or the financier Savalette de Buchelay in an Oriental dressing gown and slippers, dreamily gazing at shells and coral in his cabinet of curiosities, but he never deformed or exaggerated his sitters' features.

Each detail in his pictures counts. For example, the tailor in the village of Villers-Cotterets is shown in the main square, hat under his arm and carrying a cane. Since he does his rounds on foot, he wears dark stockings—the state of the streets allowed only those who could afford transport to wear white hose—and his wig is old-fashioned, because members of the petty bourgeoisie had to make theirs last. It is thus an exceptional visual record that Carmontelle has left us, abounding in everyday details that are not found in more official portraits—not only fashion, hairstyles, and makeup (the famous circle of rouge on each cheek that Benjamin Franklin claimed was applied with a stencil) but also furniture, crockery, worktables, sewing tables, musical instruments, games, and even the canvas covers that were placed on chair backs so that wig powder would not become encrusted in their costly upholstery.

Painting high society

Forced into the task of recording aristocratic society, Carmontelle made of it an oeuvre in its own right: a comprehensive portrait of his times, realized almost completely in profile—the manner in which historical figures had been represented in antiquity. He would never hesitate to defend his vision, and he refused to give his work to sitters since it formed part of a greater whole. When Mme du Deffand begged and insisted, he made a second portrait, but modified a detail—it is not an exact copy.

In the pages of Carmontelle's albums, the entire Orléans entourage is present—the princes and princesses, dukes and duchesses, marquises and marchionesses, counts and countesses, barons and baronesses, marshals, generals, and officers of all ranks, bishops, *abbés* and *abbesses*—all those who frequented the Palais-Royal and the family's rural seats. There they wore *habits de compagne* such as that of the warden of Saint-Cloud, M. de Mornay, slumped in an armchair in the middle of a nap, his feet up on the mantelpiece (fig. 5). And then there are those who labor: the messenger at Saint-Cloud or the gardener (whom he chose as a model in an engraving exercise he set the young prince), the milkmaid with her donkey, the ragged man hired to guard the wheat fields at Villers-Cotterets, and the queen's cobbler (a fan of tragic dramas, who is shown walking beneath the walls of Versailles, a book of plays in his hand)—men, women, and children Carmontelle encountered every day.[16]

Celebrities also appear. Diderot was captured leaning on the shoulder of his friend and colleague Grimm. As he told Mme d'Epinay: "I began to feel the cold while Carmontelle was drawing us."[17] Voltaire is portrayed adoring the famous mathematician Émilie du Chatelet;[18] Rousseau is shown at Ermenonville. We find Gribeauval who, through his innovations in firearms, would beget German losers and American victors; there's Parmentier, with ships in the background,

FIG. 5 M. de Mornay napping, by Carmontelle.

a reminder that potatoes arrived from across the Atlantic; we see Cesare Beccaria, a pioneer of modern criminal law and the first advocate for the abolition of the death penalty; and the baron d'Holbach, who announced the end of religion and the king's role as the nation's moral guardian; but there are also Necker's expansive daughter, the future Mme de Staël, and the romantic Julie de Lespinasse pining away for love in a black dress on a balcony, just as in a popular song of the time: "Anne, my sister Anne, do you see nothing on the horizon?"[19] As for Garrick, the celebrated Shakespearean actor who roused audiences to a frenzy during his Parisian sojourns, he not only inspired Diderot's *Paradoxe du comédien* but also a double portrait by "M. de Carmontelle [who] drew Garrick in a tragic attitude, and opposite this Garrick placed a comic Garrick in a doorway, who has surprised the tragic Garrick and is making fun of him."[20] Few were the famous or eminent in Enlightenment Paris who were not immortalized by Carmontelle's pencil.

This unrivaled pictorial album of life at the end of the ancien régime brings out the fads and fashions of that society. The number of portraits showing sitters at the harpsichord or plucking at a harp or a guitar proves the extent to which music was in vogue, exemplified by the flute-playing composer Frederick the Great of Prussia. The marquise de Pompadour played the breeches role of Colin in *Le Devin du village* (The village soothsayer), written and composed by Rousseau.[21] We learn in the *Mémoires secrets* that all anyone is talking about are the entertainments put on by Guimard, the celebrated actress, in her superb house in Pantin for the wits and philosophers who gathered there, and that "M. de Carmontelle wrote his collection of *Proverbes dramatiques* to be played there, which will be set to music by Laborde."[22] Among Carmontelle's numerous productions are five comedies with ariettas, six two-act operas, three one-act operas, and one *comedie-ballet* in four acts.[23] Florence Gétreau analyses Carmontelle's portraits of musicians in these pages, and provides us with an inventory of all the professionals and amateurs who composed or sang, or played instruments such as the harpsichord or flute.

Often Carmontelle sets his subjects outside: on a terrace, in a garden, or in the countryside. He depicted the duchesse de Chaulnes first as a teenage bride but later as a gardener (fig. 6). In spite of her occupation, she is dressed in virginal white; after all, she had been abandoned by her husband on her wedding night.[24] Carmontelle's humor is fully in evidence in his portrait of Mme de La Houze and Mme de Longueil taking milk from a bucket amidst a herd of cows, yet sporting sophisticated coiffures piled up high on their heads and set with ribbons and flowers. Wasn't this how the courtiers imagined country girls, since they saw them only in pictures or at the opera?

Caroline Weber, in her essay in these pages on fashion at Monceau, analyzes developments in eighteenth-century French dress and the role that the duchesse de Chartres may have played in it. In portraying the young duchess, Carmontelle had her sit up very straight in a richly gilded armchair, as befitted a direct descendant of Louis XIV. She is enveloped in a cascade of pink flounces and lace sleeve ruffles, a pouf perched on top of her high-piled, powdered, and curled hair, looking imperiously out into the park framed by the window—the very image of the wealthiest heiress in the realm (see fig. 31).

FIG. 6 The duchesse de Chaulnes as a gardener in 1771, by Carmontelle.

Carmontelle left nothing to chance when it came to his own portrait (see frontispiece). In it he is wearing an outfit in blue and red—the Orléans colors—but it is neither the Orléans *habit de campagne* for the duke's gentlemen, nor their servants' livery. He is working on an album and smiling at a subject beyond the picture frame. He has presented himself painting human nature against a backdrop of leafy nature.

The scope and quality of Carmontelle's production was appreciated by the *Correspondance littéraire, philosophique et critique* in an article probably penned by Diderot: "For a number of years now, M. de Carmontelle has been working on a collection of portraits in pencil and tempera wash. He has a particular talent for capturing the look, bearing, and spirit of the figure more than the resemblance of the features. Every day I recognize somebody in society who until then I had only seen in his albums. These figure portraits, all full-length, are realized in two hours with surprising facility. He has managed to do the portrait of all the women in Paris, with their consent. His albums, which he adds to every day, also give an idea of the variety of conditions in life: men and women of every age and station are found willy-nilly, from the dauphin de France to the floor cleaner of Saint-Cloud."[25]

Highly conscious of the value of his oeuvre as a historic record, Carmontelle tried in vain to sell it to Hugues-Adrien Joly, keeper of the king's prints, possibly through the intermediary of the financier Lallemant de Betz, who had just sold his considerable print collection to Louis XVI.[26] But the king didn't want to purchase the entire collection, so Diderot tried to interest Catherine the Great in this "collection of nine hundred portraits drawn by him, after nature, of the most respected persons of every station in life." The empress did not want to buy them all either, since they represented French rather than Russian society; rather she asked for twenty or so to be engraved for her, which they were.[27]

The artist considered his portraits to form part of a greater oeuvre whose integrity should not be jeopardized, and thus he kept his portraits. At the auction of his effects after his death, the albums were not present. They were said to have been bought by the intimate of his final years, Richard de Lédans, who, during the Empire, sold around two hundred of the watercolors to descendants of the sitters, many of the images being the only one in existence of the person represented. In 1816, after Lédans' death, the director of the *Journal des dames et des modes*, La Mésangère, acquired the remaining five hundred twenty portraits, which were subsequently sold by his heirs to the Scotsman Duff. The latter kept the portraits of English subjects, such as John Wilkes and certain actors and actresses, and sold the rest to the duc d'Aumale, the grandson of the duc de Chartres, Carmontelle's former pupil. Aumale brought them to his château, Chantilly, which is now the Musée Condé. Its statutes forbid their resale, so there it seems likely they will remain.

FIG. 7 Leopold Mozart and his children in 1764, after Carmontelle.

The image as ammunition

Carmontelle used images to reach a wide audience quickly, such as when he wanted to mock the neo-Grecian fashion of applying columns to every new structure "by publishing a design for men's and women's outfits whose parts imitate the ornaments that Greek architecture most commonly uses in the decoration of buildings."[28] It was an immediate success, and was copied. Sometimes he wanted to help: when the young Mozart first came to Paris, he was welcomed by the adolescent duc de Chartres and performed at the Palais-Royal, where he astounded the audience. Carmontelle drew the scene. Leopold Mozart afterwards wrote to his wife, who had stayed behind in Salzburg: "M. de Méchel, a copper engraver, is working in haste to engrave the very good portrait painted by M. de Carmontelle, an amateur. Wolfgang is playing the harpsichord, I'm standing behind his stool, and Nannerl is leaning with one arm on the harpsichord."[29] Used by Mozart senior to advertise his concerts and today found on countless CD and album covers, the engraving has become world famous (fig. 7).

Publicity is also what supporters of the Calas family were seeking. In Toulouse a Protestant merchant, M. Calas, had been wrongly accused of killing his son, who wished to convert to Catholicism. Voltaire took up the affair, but too late—Calas had been sentenced to a terrible death on the wheel. After his execution, the family sought to have the verdict overturned, both to restore Calas's honor and recover his impounded assets, but to do so they all had to be remanded in custody in Paris. Carmontelle visited them in their cell at the Conciergerie and made an engraving that was circulated through the network of Diderot and Grimm, the proceeds going towards the family's legal costs. Diderot's advertising copy, which appeared under the title "Subscription proposal for a tragic and moral engraving," cleverly implied that the Académie royale de peinture et de sculpture supported the project: "M. de Carmontelle, the tutor of M. le duc de Chartres, known for his skilled, witty drawings, has produced a picture that some of the greatest masters in the Royal Academy of Painting have been good enough to honor with their approval."[30] Violence emanates from the composition, in which the family, dressed in black, is gathered together in an environment entirely of stone, with just a mean metal grill for an opening. The son is bent over and unable to read because he has lost his sight in the dingy dungeon, but joy transcends the gloom as the family learns that the sentence has been overturned. The print circulated all round Europe, Voltaire placing it at the foot of his bed and sending copies to all his correspondents.

Carmontelle also portrayed Benjamin Franklin, idolized on his arrival in France, in a very meaningful way. Instead of showing him with the lightning rod he famously invented, Carmontelle depicted the statesman against a background of masts, as a reminder that Franklin convinced Louis XVI to send the French navy to aid the American insurgents. The very democratic constitution of Pennsylvania rests on the table. "Tyrants and gods are known to have yielded to him," the caption reads (fig. 8). In displaying his admiration for Franklin, Carmontelle proclaimed his political opinions for the first time.

Staging society: a living art

"It was towards the end of 1743, and we were on the banks of the Rhine, waiting for the enemy to decide to retreat so that we could do the same. The time usually spent in such a situation is extremely long, and more than ever one needs to find something to do. That's why I began *L'Étourderie réparée*."[31] Thus out of the boredom of army life, Carmontelle began to write short plays, which he called Dramatic Proverbs, and would continue to do so for half a century. Attentively observing everyone around him, he would cast them in mini-dramas, often with irony but never, as Mme de Genlis acknowledged, with malice. He even wanted them to share his enthusiasm for

FIG. 8 Benjamin Franklin in 1781, after Carmontelle.

the observation of the world in which they lived: "When acting out proverbs we are busy trying to incarnate the character in question: when we find that the character resembles a person of our acquaintance, we start observing that person more closely; this accustoms us to becoming spectators in society and coming out of our shells, something we don't always do, either through negligence or because we don't sufficiently value other people. Once we have begun to observe, how vast is the field!"[32]

The capacity for observation that guided his drawing hand also informed his quill when he described the way in which his subjects lived. He tackled the concerns of contemporary society: from the supremacy of fashion to arranged marriages; from sexual dalliances to disdain for debts. His plays are short, and their language is incisive. He has two young lords out for a stroll say to each other what he could have heard in the garden of the Palais-Royal:

> "And your [future] wife—is she pretty?"
> "My word I haven't a clue, I forgot to ask."
> "It's true that when one marries a young lady . . ."
> "It's only ever for money."
>
> ["Et ta [future] femme, est-elle jolie?"
> "Ma foi je n'en sais rien, j'ai oublié de le demander."
> "Il est vrai qu'une jeune femme qu'on épouse . . ."
> "Ce n'est jamais qu'une affaire d'argent."][33]

The tone becomes yet more brutal when he sets two bourgeois ladies in conversation—"One marries off one's children as one sells one's horse: one always claims it's the best acquisition anyone could make, when in fact all one seeks is to get rid of it and cheat the buyer."[34] Or he writes dialogue for financiers using the vocabulary of their profession: "A woman abandoned by her husband is a chattel that has come back on to the market."[35]

Collections of Carmontelle's proverbs were published in 1768, 1769, 1771, 1773, 1775, 1781, and 1786,

sometimes with fanciful titles such as *Théâtre du prince Clenerzow, Russe, traduit en Français par le baron de Blening, Saxon* (Plays by the Russian Prince Clenerzow, translated into French by the Baron of Blening) or *Conversations des gens du monde dans tous les temps de l'année* (Conversations with society figures at all times of the year). Two supplementary editions appeared after his death. But the printed versions were less successful than the actual performances, in which amateur actors often acted out their own lives and audience members would recognize themselves in the characters (fig. 9). Carmontelle would give instructions for, and sometimes even design, the sets and costumes, and would direct the plays himself.[36] The baron de Frénilly recalled that he was "inexhaustible in his inventiveness and imagination with respect to the staging of plays . . . He needed an auditorium with bleachers, sets conjuring up a forest, backdrops depicting a village or mountains, as well as costumes and arms . . . In the space of just half an hour, bales of hay held in place with stakes became tiers of seats in the amphitheater in Verona piled up for the bourgeois of Saint-Ouen . . . Giant branches cut from the trees in the garden formed both a forest setting and the wings of the theater. As for the backdrops, Carmontelle ordered that six pairs of sheets be sewn together, that cow dung be diluted, that spinach be cooked, and brick crushed . . . The following evening, the sets and backdrops—village, trees, mountain, perspective—were all in place and created an excellent effect."[37] With respect to the plays themselves, Frénilly wrote: "You were expected to learn his prose as religiously as verses by Racine . . . He caught the tone, type, style and manner of all the different classes of society with extreme realism."[38]

In the generation that followed, Stendhal would acknowledge that Carmontelle's work, which he described as "a mirror of nature," was an essential source for his own oeuvre, whereas Musset, who was directly inspired by Carmontelle's *Le Distrait* (The absent-minded man) when writing *On ne saurait penser à tout* (It's hard to remember everything) would not be honest enough to admit it. Over two hundred of Carmontelle's plays have been published, while another fifty or so await the light of day in archives.

FIG. 10

The departure of Bathilde d'Orléans, duchesse de Bourbon, the duc de Chartres's sister, 1785–90, detail of a transparency by Carmontelle.

It took more than just theater to vanquish the "boredom of the countryside" when the Orléans clan decamped to their estates. Carmontelle also organized festivities, for which weddings were a good pretext. On April 5, 1769, in the Royal Chapel at Versailles, the duc de Chartres, dressed in a coat "of fabric spangled with gold," married the pale and thin Louise-Marie-Adélaïde de Penthièvre, who almost disappeared in her dress of silver brocade. Although carefully brought up in the principles of wisdom and virtue, she had eyes only for the prince, dazzled by his elegance, his incomparable presence, his energy, and his flabbergasting gaiety. She had fallen for him the very first time she saw him dance at Versailles, where he was gazed at admiringly by the entire court.

Adélaïde was the richest heiress in France, which was the only reason for the marriage, since the duc de Penthièvre, the bride's father, was from a bastard branch of the house of Bourbon. At the ceremony, presided over by Louis XV, Orléans was in the front row while Penthièvre was relegated to the last. A presentation tour was organized for the newlyweds through all the ducal estates. In Villers-Cotterets, during that beautiful summer, Carmontelle put on a series of entertainments: plays, mime performances extolling the virtues of the Orléans, banquets, balls, fireworks, and village dances, the latter involving young farm girls who were admired by the famous Tronchin, the family doctor: "They seemed to me too pretty not to have been chosen from a radius of several leagues." The princes and their guests—who included the Duke of York, son of the king of England, and the prince de Bauffremont—boasted of this "bucolic festivity in the manner of Carmontelle." Of the frenzied celebrations, which he deemed put his health at risk, Tronchin wrote to his daughter: "The princes are like weathervanes that only stop once they're completely rusted up."[39]

Continuing the celebrations, in mid-May the entire company decamped to the estate of their cousins in L'Isle Adam, before joining the festivities put on by Bathilde d'Orléans for her brother at the Château d'Issy, which had been lent by their uncle, the prince de Conti. All of Carmontelle's imagination was needed here, for the site was rather unfavorable. He created huge, painted-canvas backdrops that conjured up the Tuscan countryside—backcloths that were lit at night from behind by tall torches, like giant transparencies. A few months later he devised equally splendid festivities for the wedding of Princess Bathilde to the Condé heir and was soon sought out by other courtiers in need of lavish and unusual entertainments (fig. 10). In 1771 the *Correspondance* asserted that "he has taste and is one of the creators of society festivities who is most often employed in Paris."[40]

In the service of the house of Orléans, Carmontelle became a star who shined brightly in fashionable society, not as one of its members—his origins were too modest for that—but as an artist and artisan of pleasure. "He dined everywhere and was a parasite nowhere. He amused everybody as a friend who would oblige: he was an artist who charged nothing and a man of high society," recalled the baron de Frénilly, who added, very significantly, "on foot."[41] Indeed Carmontelle had no money, but he wouldn't condescend to ask for a lift. He dominated the courtiers; wasn't he able to defeat their mortal enemy—boredom?

MONCEAU

At the Palais-Royal the duc de Chartres (fig. 11) benefited from pretty private apartments with direct access to the garden. But the origin of his plans for Monceau can perhaps be found in a family situation that made them less agreeable to him. At his son's request, Orléans had dropped his mistress, the actress Marquise; Chartres considered such a liaison beneath the dignity of a prince of the blood. Although the actress was now raising their three children on an estate he had given her, Orléans had become infatuated with the marquise de Montesson and planned to marry her. Chartres, however, disliked Montesson as much as her predecessor. He avoided her at the Palais-Royal and clearly had no intention of putting up with her at Châteu du Raincy, east of Paris, an estate just purchased by his father that came with vast hunting grounds. But he had no other country haven near the capital. It was time for him to have his own estate.

A dreary plain

The hamlet of Monceau (or Mousseaux) was located on the road leading from Paris to Saint-Ouen and the valley of Montmorency. A retreat from the city that was much appreciated by Parisians, it consisted at the time of a few houses and farms. Was the choice of this site for Chartres, who was still in "the flower of youth," as Carmontelle would later put it, a continuation of paternal plans drawn up ten years earlier before the death of his mother? In 1754 Orléans had purchased the hôtel Buizette, in the faubourg du Roule on the road to Monceau, but he sold it after his wife's death in 1759.[42]

The road from the Palais-Royal to Monceau passed through the faubourgs Saint-Honoré and du Roule before ascending a hill that offered sweeping views over the capital. Below were the roofs of the sumptuous new mansion built by Soufflot for Mme de Pompadour's brother, the marquis de Marigny, director of the king's buildings, and next door were the king's plant nurseries. The site fulfilled Dr. Tronchin's recommendations. The well-known doctor convinced enlightened Parisians that they were breathing stale air, harmful to the lungs, in the center of the city, prompting them to head for windmill-topped hills around the capital.

The plain around Monceau was monotonous but was praised for its excellent hunting. Chartres, an excellent shot who was skilled at hunting to hounds, could expect to enjoy the thrill of the chase. Furthermore, a new bridge had been built over the Seine at Neuilly, which allowed Saint-Germain and Versailles to be reached much more quickly—an advantage for a prince of the blood with frequent state duties. Perhaps even more important was Monceau's relative proximity to the Palais-Royal, as Gabriel Wick points out in his essay in this volume. Wick interprets the garden as part of a public-relations strategy intended to deliver the duke, who desired

FIG. 11 The duc de Chartres in the *habit de Saint-Cloud*, by Carmontelle.

a greater role in affairs of state, from the political sidelines to which his status as a prince of the blood condemned him.

On December 19, 1769, at the Palais-Royal, the duc de Chartres signed a lease with Louis-Marie Colignon, architect to the king and building contractor. The document gave Chartres the lifelong usufruct of a yet-to-be-built villa surrounded by a garden on the heights dominating the faubourg du Roule. Annexed to the lease were plans of its proposed interior layout and the plant species designated for the garden. Elevation drawings showed the materials to be used on the house's exterior. But would Chartres be content with a villa in the middle of a small, formal French garden? Apparently not. The *Mémoires secrets* reveal that praise for Simon-Gabriel Boutin's large and lavish garden, Tivoli, had "inspired the duc de Chartres to create a similar garden of his own, but even more magnificent and worthy of His Highness."[43] The prince turned to his tutor, portraitist, author, scenographer, and events organizer. Carmontelle must create one of his enchantments at Monceau!

(75)

The space available at Monceau was so constricted that Carmontelle chose not to wall off the property; rather, he dug ha-has that allowed the landscape to the north and south to be visually integrated into the garden. Probably he discovered the idea of "borrowing the surroundings" in Japanese scrolls that had been bought by the Regent. He may have drawn also on Engelbert Kämpfer's *Histoire naturelle, civile et ecclésiastique de l'empire du Japon*, which was translated from English and published in 1729, and Sir William Chambers's *A Dissertation on Oriental Gardening*, published in a French translation in 1772.

Chartres continued to acquire neighboring plots in a bid to enlarge Monceau, but it would never reach the size of his father's park at Le Raincy, let alone that of many English estates. Elsewhere in this volume David L. Hays analyzes the park's creation meticulously, recording every successive acquisition of land and the evolutions in the plans engendered by each enlargement.

An Enlightenment garden

There remained the task of designing the garden's interior and creating surprises for visitors on the unremarkable terrain, as Susan Taylor-Leduc describes in her essay in this volume. At Monceau, Carmontelle proposed a fantasy that would turn the visitors into both spectators and actors: a promenade of around two hours led the prince and his entourage from one scene to another, with games and distractions offered along the way. The writings of Hellenist Johann Joachim Winckelmann, the Count of Caylus's *Recueil des Antiquités*, and the discovery of Herculaneum and Pompeii had inspired French architects with visions of the ancient world, so Carmontelle bowed to fashion and created three scenes evocative of antiquity: a ruined temple of Mars, an open-roofed temple in the form of a rotunda, and a colonnaded Naumachia that harked back to the staged naval battles beloved by the Romans.[44] He also designed three "oriental" scenes, with features that he dubbed the Tartar and Turkish Tents, and the Minaret. The last served as a belvedere which offered an expansive, 360-degree view of Paris, her monuments, and the surrounding region. From the Minaret Monceau appeared to be at the center of everything. It and the garden's other orientalist trappings

FIG. 12 *Mme de Pompadour as Friendship*, 1753, by Pigalle.

were manifestations of current cultural preoccupations. Voltaire's *Zaïre*, Rameau's *Les Indes galantes*, and *turqueries* of every sort were all the rage, and people drank coffee while coiffed in turbans and shod in Turkish slippers. Liotard portrayed Louis XV's daughter, Madame Adélaïde, as an "Orientale" languidly reading a book on a sofa. Carmontelle also designed a pagoda-style structure to shelter a brass-ring carousel evoking far-off China, whose porcelain and lacquerware were so appreciated in France by Mme de Pompadour that royal manufactures were created to copy them. During a magical night of festivities that Carmontelle organized at Saint-Cloud, Chartres and twenty-three of his friends, dressed as Chinamen in cloth of gold, danced a quadrille on floating platforms in the château's torch-lit pool.

Nature was now becoming fashionable. Civilizations distant in space and time having been honored, Carmontelle filled the rest of the park with bucolic scenes: a farm with an inn where visitors could obtain light refreshments; the Horse Chestnut Chamber where strollers could rest in the shade; two French pavilions surrounded by flower gardens; and a wooden bridge suggestive of the Rousseauian ideal of rusticity or nature. Carmontelle also added an Italian-style vineyard where visitors could pick grapes, and a Dutch windmill operating a steam pump that kept Monceau's waters flowing to prevent stagnation.[45] The hydraulic circuit, which was animated by a water mill and a waterfall next to a ruined castle, formed a sort of backbone to the garden. The castle, which symbolized the feudalism of medieval France, and a wood dotted with tombs were intended to remind the princes of the ephemeral nature of their power.

Of course Carmontelle did not forget the horticultural aspect of Monceau. Georges-Louis Leclerc, comte de Buffon, the preeminent French scientist of the day, was a major force in popularizing the study of nature. Everyone flocked to hear him at Paris's natural-history museum. Carmontelle drew an extraordinary portrait of Buffon looking at the plants and animals he studied, his instruments behind him. At Monceau, Carmontelle would bring coherence to the illusion by choosing plant species linked to the scenes being represented. As Elizabeth Hyde confirms in her essay in the following pages, choosing the right plant was as important as constructing the right folly or *fabrique*.

A collector's pleasure

It naturally went without saying that a prince of Orléans, who counted among his ancestors two great collectors, would be surrounded by works of art wherever he resided, and Carmontelle was at pains to provide sculpture for the garden. If we follow the itinerary he mapped for the garden, we arrive first at one of his mystifications: "In the middle [of the temple] there was a statue of Mars which was too disfigured; it has been replaced by one of Perseus, which is antique. The temple has none the less kept the name of Temple of Mars" (*Garden at Monceau*, 19). Three other antique sculptures appear as one continues along the route: Bacchus is in the middle of the Italian vineyard; Mercury, the god of travel, is found at the crossing of the paths that run through the wood; and beyond the wood we encounter Meleager, but without a wild boar in sight. Fashionable contemporary artists were asked to contribute works also, Jean-Baptiste Pigalle providing a statue of Hymen (the god of marriage) and its pendant, Friendship (fig. 12). The latter, commissioned by Mme de Pompadour for the park at Bellevue, had been sold by Marigny on his sister's death, bought back by the sculptor, and then resold to Orléans, who gave it to his son.[46] Then there was Jean-Antoine Houdon who, in a tribute to a renewed Mediterranean antiquity, created an exceptional group sculpture that defied the rules of the Académie royale because of its diverse materials. At the center of a pool, the sculptor placed a seated bather carved from white marble. In striking contrast, a standing African servant made of lead poured water over her from a golden ewer (fig. 13).[47] The water from the ewer fell into a basin that drained into a large pool of the Naumachia.

More sculpture could be found inside Monceau's large pyramid, including Egyptian heads supporting an entablature and a fountain figure whose breasts jetted water that fell into an antique marble basin. There was also a statue of Henri IV, a Chartres ancestor and very popular king, which reinforced the duke's public persona. Finally, a copy by Edme Bouchardon of an antique faun from the Villa Barberini was placed in a niche and was the last thing that visitors saw (fig. 14). The young Bouchardon realized this splendid marble replica at the French

FIG. 13 Houdon's *Bather*, 1782.

Academy in Rome in 1726–30, and Louis XV gave it to Mme de Pompadour's brother, the marquis de Marigny. In 1773 Marigny first rented and then sold his house in Le Roule to the duc de Lauzun, a close friend of Chartres; at this point the statue was moved to Monceau.[48]

Monceau as museum

Carmontelle would only accept one adjective to describe his garden, "picturesque" (in the original sense of the word as derived from "painter"). Others before him had pursued the goal of garden as painting. For example, the banker Henry Hoare, whose garden at Stourhead was one of the finest English parks, had learned many lessons from artists working in Rome. In a remark recorded by Joseph Spence, Hoare wrote: "The green should be ranged together in large masses as the shades are in painting: to contrast the *dark* masses with *light* ones, and to relieve each dark mass itself with a little sprinkling of lighter greens here and there."[49] And a letter to his daughter reveals the extent to which he was inspired by Claude Lorrain and Claude's pupil, Gaspard Dughet: "the View of the Bridge, Village & Church altogether will be a Charm[in]g Gasp[ar]d picture."[50]

FIG. 14 Bouchardon's 1730 replica of the Barberini Faun.

Carmontelle perhaps dreamt of becoming a great painter. In 1770 he was well over fifty. While Diderot acknowledged the acuity of his eye—"Any man who knows how to draw only like our friend Carmontelle, without having more verve than him, has but to cross the tax barriers at five o'clock in the evening or at nine o'clock in the morning to find subjects for a thousand pictures"—he nonetheless added this deadly appraisal: "But these pictures will only arouse curiosity in Moscow"![51] Carmontelle knew it.

Instead, at Monceau he created for his pupil a series of three-dimensional scenes that formed an open-air museum of "*all times and all places*." The *Encyclopédie* defined the term "museum," an institution invented in the eighteenth century, as "any place where things that have an immediate relationship with the arts and the muses are kept." The galleries of the Palais-Royal offered up the beauty of their masterpieces to the admiring gaze; Monceau presented its treasures in a landscape. In so doing his garden fulfilled the mission of edification that underlay the whole Enlightenment approach—the idea of discovering other cultures in both time and space—an extraordinary goal that disappeared with the Romantic withdrawal into nationalist patriotism. Just as the members of the house of Orléans since the time of the Regent kept the Palais-Royal open to whomever asked to visit it, so Monceau could be accessed by the public. Documents at the Archives de Paris reveal that Alexandre Lenoir, saviour of numerous works of art, hoped to use Monceau for his museum of monuments during the Revolution.[52]

Amusing a prince

The garden of illusions was also a stage set, and the director, as was his wont, incorporated his spectators into the scenario. He dressed "the persons who serve the Prince in different costumes, when he dines or sups in his pavilion, or strolls in his Garden" (*Garden at Monceau*, 18). If the prince and his guests, playing the role of members of a desert caravan, were riding the dromedary, then naturally the servants were disguised as camel drivers!

A small Gothic building, at the time a symbol of the obscurantism that the Enlightenment set out to illuminate, contained a chemistry laboratory, Chartres having decided, like his great-grandfather the Regent, to cultivate this fashionable science. Arguably more recreational was the frightfully swanky billiard room, which Carmontelle hid beneath one of the Turkish Tents. The sports-loving prince could also bet with his friends when challenging them to a shooting match. The target was a popinjay, a colored wooden bird set up at the top of a pole. A bell behind it rang each time the target was hit. Louis-Racine de Monville—the dandy son of a *fermier général* (a financier serving as a tax collector) who was not only the ladies' favorite but also an excellent rider—was almost unbeatable at the popinjay, and it was to try to better him that he was invited.

On the pagoda-like, man-powered Carousel, the prince and his friends—the men perched on phantasmagoric animals, the ladies supported by male mannequins—amused themselves in attempting to detach a set of brass rings with slender sticks, an entertainment that was not without its erotic side. The Carousel, which had merrily jingling bells attached to its pavilion, was set up on a tiny man-made island with drawbridges.

Again for the pleasure of the prince and his guests, Carmontelle painted a see-through spring landscape on the ten windows of a gallery in the villa at Monceau. In the throes of winter he "was able to make spectators believe they were in the finest season of the year, and even in summer the landscape of these paintings harmonized perfectly with the real objects that, through one of those windows, nature offered to the eye" (fig. 15).[53]

FIG. 15 Detail of an illustration from Carmontelle's 1794 "Mémoire sur les tableaux transparents du citoyen Carmontelle."

The garden as a militant work of art?

A little round temple of white marble contained a statue representing one of the women Achilles lived among when he was hiding at the court of King Lycomedes. Two other neoclassical structures were constructed as ruins: the Roman Temple of Mars and the Naumachia. Perhaps their condition was meant to foreshadow the inevitable end of the British Empire, which was often compared to its Roman predecessor. In contrast, the intact temple could suggest the resistance of small, Greek-style republics such as Holland or even the future republic desired by the rebellious English colonies in America. The three classically inspired follies as a group paid homage to Winckelmann, whose recent murder in Trieste had caused a sensation.

The Dutch Windmill was built at the highest point of the garden, both to catch the strongest breezes and to harness gravity in the system of water circulation. This prominent position honored Holland, an industrious society where civic and religious freedoms were guaranteed—the Calas affair had not been forgotten. The park's second highest spot was used to evoke Islam with the Minaret. In the *Mercure de France*, Voltaire acknowledged that "as in the second century of Mohammed, the Christians of the West had to seek their education from the Arabs," where they discovered not only the texts of antique Greece but also astronomy, medicine and chemistry.[54]

(79)

Some see Masonic influence in the garden at Monceau—and Chartres was indeed appointed grand master of the Grand Orient at the end of 1773 when the chapter was created by the duc de Luxembourg, who was seeking to put an end to infighting in French Masonry. The nomination of the young duc de Chartres, first prince of the blood, was intended to calm quarrels to which he was entirely external. He had not been a Mason previously, nor had his father or his father-in-law, Penthièvre or Carmontelle, for that matter. The duc de Luxembourg could thus easily establish his authority, since it would have been unrealistic to entrust effective power to a noninitiate. The duc de Lauzun held the post of secretary, and it was perhaps he who suggested that Chartres be appointed.

While there are traces of a Masonic lodge at Monceau—the grand master obviously would require one—it doesn't seem that any activity ever took place there. Moreover, if Monceau was designed as a Masonic park, why did Chartres commission Thomas Blaikie to remodel it entirely as of 1782? And why did his successor, Cambacérès, who was by turns second consul and arch-chancellor of the Empire as well as an active and cognizant Freemason and the architect of the revival of Masonry after the Revolution, himself not hesitate to undertake profound modifications?[55]

Monceau à la mode

At the beginning, Monceau was a place of promenade for the prince. His wife then asked him to organize festivities there for herself and her friends, which he did willingly. Soon they brought their son there, the duc de Valois, born in the fall of 1773, because the air on the plain was healthier for the child.[56] The duchess came often, and the baronne d'Oberkirch recalled the supper organized there for the first French visit of the Count of the North and his wife, a.k.a. the Czarevich Paul and the Grand Duchess Maria Feodorovna, in June of 1782. Arriving from Saint-Cloud, the baroness, who was accompanying the royals, carried a magnificent bouquet of rare flowers, which the comte de Mornay sent every morning to the duchesse de Chartres, who adored them. The baroness confided in her memoirs that "Madame la duchesse de Chartres was most particular about decorum . . . She loved M. le duc de Chartres with one of those affections that can resist everything . . . We retired very late from Monceau, the supper having run on for a very long time. Everyone was so amiable at this court."[57]

The streets around Monceau were baptized "rue de Chartres" and "rue de Valois," and the prince issued numerous invitations to his garden. Everybody gave their opinion on it, the duc de Croÿ seeing in it nothing less than the "rich tone of China."[58] Saint Aubin and Hubert Robert came to paint at Monceau,[59] while Rousseau went "without having to ask permission"; Mme de Genlis gave him the key.[60] Although unwell, Franklin had himself carried there in his sedan chair, while the abbé Delille composed poetry in the garden. Monceau had become famous; Carmontelle had satisfied all his master's desires.

Carmontelle was now much in demand, as the baron de Frénilly recalled: "He designed and planted rather extraordinary gardens since they weren't French, and he flew into a rage if anyone called them English gardens. He planted my father's garden at Saint-Ouen and the famous garden at Mousseaux, on the wall of which he wrote: *This is not an English garden.*" Frénilly also added a remarkable compliment: "One would more easily have done without Le Nôtre under Louis XIV than without Carmontelle in those days."[61]

The dream and illusion made solid at Monceau inspired the queen, who commissioned her architect Mique to copy the brass-ring Carousel as well as the Farm (fig. 16), which became her famous hamlet at Versailles. At Bagatelle her young brother-in-law, the comte d'Artois, also wanted a few follies, even though he had asked the Scottish gardener Thomas Blaikie, a fervent exponent of the naturalistic mode that reigned supreme across the Channel, to create the setting for his extravagances. Monville threw himself into the

creation of a park at Retz, which he christened "the Desert"—the name alone is a whole program—whose delights included a pyramid, a circular temple dedicated to Pan (logical for a man who was also a musician), a Tartar tent, a Chinese pavilion, a Gothic ruin, a farm, and a residence in the form of a ruined column. Bélanger and Hubert Robert toured Monceau before creating a garden at Méréville that was made for dreaming. Gardens were all the rage across Europe where princes, bankers, merchants, and the newly rich were building country residences. While few dared take up the T-square and the compass, many were those who displayed a passionate interest in the settings for their jewels, laboring alone or with the help of a gardener. Gardening came to be regarded as an art in its own right. Mozart set it to music: in *Die Entführung aus dem Serail*, the valet, whose master has passed himself off as an architect, wins his freedom from the pasha by claiming to be a gardener.

But passion soon gave way to theory; Carmontelle's originality and inventiveness brought him a success that exasperated some. In 1776 Jean-Marie Morel, who designed the gardens at L'Isle-Adam for the prince de Conti, published a *Théorie des jardins* in which he violently criticized Monceau. Announcing the publication of Morel's book, the *Correspondance littéraire* offered Carmontelle the opportunity to defend himself, explaining that, given the lavish praise heaped on Morel's voluminous text, the editors had sought out the view of a specialist who "has long dealt with the art of gardens." It is interesting to note that those behind the *Correspondance* had at times been critical of Carmontelle's work, but they were sufficiently appreciative of Monceau not only to offer him a platform in their publication but also to state their own admiration. "It is M. de Carmontelle who drew all the plans for the garden of M. le duc de Chartres, a garden of which M. Morel speaks with only the most regal disdain, but which is nonetheless the work of a highly fertile and ingenious imagination; and one can judge it by the drawings of it that M. de Carmontelle produced himself and which form a suite of very varied and picturesque landscapes that the artist's magic alone managed to produce on a rather thankless terrain, which he was not even always able to organize to his liking . . . The artist and the amateur will perhaps find there more resources and invention than in all our learned theories."[62]

Carmontelle wasted no time in exercising this right to reply. In seven pages he mocks Morel's verbose theories and ridicules his pedantic tone: "You ask me, Monsieur, what I think of the volume *La Théorie des jardins*; I found in it a fine description of the countryside; the countryside is everywhere, it's true, but one hadn't yet seen that in *La Théorie des jardins*, and it gives the publication a philosophical air that is marvelously convincing. Eh, Monsieur! who taught you so much?" he snickers. He then advises the author to learn to make gardens rather than explaining that "gardens don't come out well everywhere," particularly on arid plains. He sneers at the statement that the design should not determine everything, and reminds the reader that one can and must make models in order to judge correctly. He even specifies the rules of perspective: "When one looks at [a section of] countryside, a park, a garden, a coppice, etc. outside of its boundaries, everything will appear small in comparison to the immensity that surrounds it; but it is within that one walks, and objects seen close up are then far larger than this immensity."[63] Finally, in response to Morel's regret at finding "every century and every part of the world" at Monceau, Carmontelle sarcastically retorts: "The author would no doubt like all the pictures in a house, being considered as holes made in the walls, to offer up to the gaze only that which can be seen through the windows so as not to revolt the senses and give a foothold to implausibility; in consequence one could not have, even in painting, the finest monuments of Italy, or Flemish and Dutch architectural follies, and certainly not any seascapes, for the sole reason that one lived on dry land!"[64]

The garden in print

Carmontelle feared for the future of Monceau, making his attack on Morel all the more severe. Chartres seemed to have fallen under the influence of another silver-tongued roué, Marc-René d'Argenson, the marquis de Voyer, son of a secretary of state in the Department of War as well as the lavish director of the royal stud farms, who proclaimed in the society salons "our sole pleasure must determine all our actions." Talleyrand described him thus: "He was the leader of all the corrupt men . . . He denied the existence of morality, maintaining that among men of wit it was merely a word."

Talleyrand recognized Voyer's pernicious influence on the young Chartres: "It is from this second education dispensed at an age when men are disciples of all that surrounds them that the corruption of M. le duc d'Orléans [Chartres's title from 1785 onwards] can be dated."[65] Indeed from 1780 the duke and his friends, who sought only pleasure, little by little transformed Monceau into a place of debauchery. The prince de Ligne confessed that "not liking to be left behind in anything, I often drank two bottles of champagne when it was fashionable in the company of M. le duc de Chartres."[66] It was from this moment—after Carmontelle had launched his creation into the world—that Monceau became a center of libertinism. But the seeds were there in the garden's design, as Joseph Disponzio suggests in his essay in this volume.

Although Voyer recommends in his *Manuel des Châteaux* that Carmontelle's works be read, the designer worried nonetheless: Might not the fickle duc de Chartres ask Thomas Blaikie to transform Monceau in the English style, as his cousin, the comte d'Artois, had redone the grounds of Bagatelle? Carmontelle decided to preserve his creation for posterity with a publication in which he would develop his ideas with respect to gardens; describe his most famous creation; and depict it in a series of engraved plates.

The prospectus

Faced with the costs of publication, Carmontelle organized a subscription and published a prospectus that deserves to be read attentively. From the outset, Carmontelle takes aim at Monceau's critics, retorting that "this sort of Garden can . . . not be judged like ordinary Gardens" (prospectus, 7), and boldly using the example of Versailles to counter the arguments of those who said the park was just a muddled jumble: "If one did not know what the Gardens of Versailles were like before they were replanted, would one believe that they are Le Nostre's masterpiece, seeing them as they are now, where all that adorns them can be seen at once?" (prospectus, 7).[67] Then he reminds readers it would take time for the vegetation to grow and create distinct spaces. When taking a dig at the fashion for English gardens which were supposedly closer to nature, he adopts a sarcastic tone, speaking of "a Nation which, when making natural Gardens, runs the roller over all the lawns & spoils nature by showing everywhere the prissy art of the unimaginative Gardener" (prospectus, 7)—and cites the *Dissertation on Oriental Gardening* by "M. Chambert" (Chambers) to back up his argument. Then, having paid homage to a man skilled in the "art of creating modern Gardens," Sir Thomas Whately, he finally states his credo: "one will produce a very handsome Garden, especially by never forgetting to contemplate nature's most beautiful effects, which one should study as a Painter, by exploring them in the countryside, pencil in hand" (prospectus, 8). This then was the goal—a painter's garden, a picturesque creation.

Next he explains: "In the Garden at Monceau, everything had to be created; we took advantage only of the views of the surroundings, which we appropriated by enclosing the Garden with nothing but ditches, which makes it impossible to estimate its true extent" (prospectus, 8). The garden was designed for a two-hour promenade where "the different scenes that appear at each moment are the source of all the pleasure" (prospectus, 8). He then announces the seventeen views that make up the publication along with a detailed plan that allows readers to locate each one. Launched at the end of 1778, the subscription was apparently successful since the album was published over the course of 1779.

A handsome edition

The volume comprised a title page, a foreword, ten pages of text, and eighteen plates. The text allowed Carmontelle to specify the circumstances of the commission, describe the original state of the terrain at Monceau, and detail the design he drew up for the prince and his friends. "We do not have the pretension of offering here a theory or precepts; it would be ridiculous" (*Garden at Monceau*, foreword). He gives a nod to Whately, whose *Observations on Modern Gardening* had been translated by Latapie in 1771 as *L'Art de former les jardins modernes* (subtitled *L'Art des jardins anglais*), and the duc d'Harcourt, whose unpublished treatise he admired, but says that he would never dare compare himself to them.[68] Since his modesty prevented him from considering the creation of gardens as a great art—only history painting enjoyed that status at the time—one had to invent and surprise: "The goal of the Arts in great things is to force the soul into admiration, to carry off, bind, & dominate it through the power that real beauty always has the right to exercise over them. In those that are but agreeable, the soul must continually be moved, interested, & amused by the charm of images, because it cannot do without them" (*Garden at Monceau*, 13).

He humorously analyzes the attitude of the fashionable with respect to gardens and nature: "What are the country charms that are cited in France? Pure air & freedom; & one almost never enjoys either one or the other. But would they be sufficient for us? We would find few charms in the countryside without those of society" (*Garden at Monceau*, 13). Here is the heart of the matter: the French, or rather his clients specifically, seek only to vanquish the boredom engendered by their idleness. They are incapable of looking after a farm; the only country people they know are to be found in paintings by Boucher, in the pastoral novel *L'Astrée* or on the stage at the opera; they avoid the sun, the damp of the woods, and the persecution of insects; and they find pleasure only in festivities in the company of their own sort. Carmontelle champions this particularity of the French elite and opposes it to English attitudes, which, he argues, should not be imposed outside British shores: "We love this happy freedom that produces new & piquant effects; what is more we have our own ideas, tastes, & customs; if they depend on our climate, any attempt to have us adopt those of our neighbors will be in vain; when we diverge from our old principles, we will invent new ones that will belong only to us: while one may divert the waters of a spring, one cannot alter their quality" (*Garden at Monceau*, 15). He argues against the monotonous uniform green of vast English lawns which "would be too sorrowful for our soul" (*Garden at Monceau*, 15) and attacks the fashion for English-style white fences because "yellow & red agree better with the greenery of the different trees." He pleads in favor of fantasy, far from the ideals of Le Nôtre: "Picturesque Gardens need fortuitous negligences, which belong only to taste" (*Garden at Monceau*, 15).

Horace Walpole's ironic comment on the French garden—"When a Frenchman reads of the garden of Eden, I do not doubt but he concludes it was something approaching to that of Versailles"[69]—drew return fire from Carmontelle: "If we felt the desire to imitate English Gardens, it was only to escape the monotony of ours" (*Garden at Monceau*, 15). He takes the precaution of stating that Monceau was not intended as a model, although he adds sarcastically that many are incapable of appreciating it: "We have masters to help us learn to talk, dance, & sing, &c., & yet we do not think of learning to see" (*Garden at Monceau*, 13). Only painters can teach that: "they make you perceive the gradations of linear & aerial perspective, they unveil the secrets of space; they will reveal to you the diversity of all the color tones, their relationships, & their harmony. Without such knowledge, we see only the skeleton of nature" (*Garden at Monceau*, 13). His words can thus be read in the context of his role as teacher to the prince; learning to see is part of the process of emerging from the state of ignorance.

Carmontelle laments the frivolity of the garden's visitors who, when they climb the Minaret, "count the bell towers & judge the view only by the remoteness of the objects they perceive" (*Garden at Monceau*, 14). This, he explains, is a function of their incapacity to observe thoughtfully: "What they call a fine view is but a horizon of which they see only the points where it terminates" (*Garden at Monceau*, 14). Despite everything, the inattentive prince who had given him his entire confidence, is allowed a word of gratitude: "when one is lucky enough that he to whom the Garden belongs . . . steadfastly refuses to have in it what one finds in all the others, then this rule makes you seek out new ideas, combine them, & multiply them . . . Let us use this freedom to enchant, to amuse, & to arouse interest" (*Garden at Monceau*, 14). He adds: "This is what those who came to see the Garden at Monceau were hoping to find; because they said when there was not much, *I would be just as happy to walk in the countryside*" (*Garden at Monceau*, 14–15).

On his plan of Monceau (plate I) Carmontelle marked different points with letters of the alphabet keyed to descriptions of the plates. They are short and precise, and emphasize the technical, and especially hydraulic, problems that had to be overcome. The plates form the second half of the book. Carmontelle, who uses the signature "L. C. de Carmontelle" (Louis Carrogis de Carmontelle), commissioned nine engravers to produce them from his drawings (Bertrand, Cauché, LeRoy, Deni, Michel, L'Epine, Michaud, Croutelle, and Colibon).[70] Each view is animated by people (acquaintances of the prince or servants in costume) and animals (cows, sheep, and of course the dromedary). The first view (plate II) sets the garden in its environment by showing its integration with the surrounding landscape. The windmills that punctuated the heights surrounding Paris, especially on "the hill of Montmartre," are visible in the background. Each of the following plates shows one of the scenes that made up the garden. Carmontelle alternates near and distant views, and centered compositions with those in which the "objects" are placed to one side to give more importance to the overall vista. The final view (plate XVIII), titled *Salle des Marronniers* (Horse Chestnut Chamber), is an homage to Bouchardon's *Sleeping Faun*. To the right, a man is pointing out sculpture to a group of three ladies who are discussing it. At left another two ladies, one gesturing at it with her fan, seem to be awaiting the verdict of a second man, who is portrayed in a pensive attitude. Was this a nod to Carmontelle's prince?

Monceau did indeed end up in the hands of Blaikie, who described it thus in his diary: "those Gardens which had been done by Mr. Carmontelle Architect to the Duke de Chartres at very great Expense where there was Monuments of all sorts, Countries and ages but placed in such a manner that from every part there was a confused Landskipe, for there was adjoining Chinese & Gothic buildings, Egyptian Pyramids joined to Italian vineyards, the winter Garden adjoining to the Hothouses, was more Beautiful than Elegant; the whole was a Small confusion of many things joined together without any great natural Plan, the walks Serpenting and turning without reason which is the fault of most of those gardens done without taste or reason; after I changed most of those Gardens and destroyed most of those walks which I thought unnecessary or unnatural."[71] In resentment Carmontelle sought vengeance by portraying Blaikie as a ragged gardener harnessed to a wheelbarrow, with a high-towered medieval castle in the background—a reminder that the Scotsman required every large park to have one.[72] Next, Carmontelle wrote a dramatic proverb, *Le Jardin anglais* (The English garden), which mocked the fashionable by making them laugh at the way they themselves chased after the latest fads. The play ends with a chambermaid entering to make an announcement to her mistress who is receiving guests. Among them is a chevalier, just back from England:

> The Chambermaid: "Madame, the windmill is turning, the cave is frightfully dark, and the ruin has been completed."
> The Chevalier: "What do I hear? Where am I? What language is this?"
> The Baron: "It's the language of English gardens."
> The Chevalier: "Of English gardens?"
> The Baron: "Yes, the marchioness has just made one in which we are now, with caves, ruins, and tombs."
> The Chevalier: "Tombs?"
>
> [Femme de Chambre: "Madame, le moulin tourne, la caverne est bien sombre et la ruine est achevée."
> Le Chevalier: "Qu'entends-je? Où suis-je? Quel language?"
> Le Baron: "C'est celui des jardins anglais."
> Le Chevalier: "Des jardins anglais?"
> Le Baron: "Oui, la marquise en a fait un où nous sommes actuellement, où il y des cavernes, des ruines et des tombeaux."
> Le Chevalier: "Des tombeaux?"][73]

So ends the proverb, just as Carmontelle's long learning curve in the company of the duc de Chartres came to end. "When he saw the scheming of his detestable prince," Frénilly recounts, "Carmontelle nobly quit the post of tutor from which he earned his living."[74]

THE END OF THE ENLIGHTENMENT

Anticipating cinema

But the duchesse de Chartres asked Carmontelle to remain in her employ, which he did willingly. He took up his brushes once again at the request of Mme de Genlis, who had become governess to the Orléans children, and taught them "to see." His method was still pertinent, if we judge by the sketchbooks of the prince Antoine and his sister, the princess Adélaïde.[75] For the duchess, ever more deserted by her fickle husband, but just as much in love with him, Carmontelle invented his transparencies, an ancestor of cinema. He would take images similar to those he had realized on the windows of the Monceau gallery, paint them on strips of translucent paper, and set them in motion.

FIG. 17 Illustration from Carmontelle's 1794 "Mémoire sur les tableaux transparents du citoyen Carmontelle."

The innovation was probably linked to his discovery of the Japanese painted scrolls known as *emaki-mono*. Popular during the Heian era (late-eighth through late-twelfth century), *emakimono* are continuous landscapes that tell the stories of humans or animals. Often painted in monochrome ink, they range from twenty to forty centimeters high, can reach twenty meters in length, and are made from sheets of paper that have been glued together. They are stored in wooden boxes, rolled around batons made of valuable woods.[76] Carmontelle, a man of the theater, improved the technique so that his scrolls could be seen by several people at once and have a more magical effect. He would write a script, decide on each view, and then design and paint the scenes on glued-together sheets of high-quality translucent paper, each measuring around seventy centimeters wide and thirty to forty-five centimeters high (the height increased as the manufacturer, the Englishman Whatman, improved his techniques). The consecutively painted views formed a "film" of up to forty-five meters that was attached at each end to wooden rollers enclosed in black boxes. When the rollers were turned, one scene followed the next without interruption (fig. 17).[77] Staging stories set in the parks and countryside around Paris, Carmontelle had created a theatre that no longer required cajoling the undisciplined actors who were the prince's friends.

These transparencies were yet another contribution to his depiction of the society in which he lived. "Constancy and application were required to capture all the sites," according to a posthumous catalogue. Carmontelle also would invent dialogue and, as was the practice with the first silent movies, included a musical accompaniment supplied by a harp or harpsichord or even one of the new pianoforti. "Society persons appeared in action" in these productions, recalled Frénilly.[78] The duc de Charost was particularly fascinated by the new art form and "had [Carmontelle] accept a pension of four thousand francs as the price for one of his transparencies."[79]

In 1786 Carmontelle published *Conversations des gens du monde dans tous les temps de l'année*, whose foreword sets an acid tone: "We believe ourselves obliged to warn those who read this work that they will find in it nothing new, and that we have collected in it only that which we hear said every day: the author's goal is therefore not to educate but on the contrary to teach foreigners how to speak without actually saying anything."[80] Mme de Staël sent part of the text to the king of Sweden: "I'm sending Y[our] M[ajesty] a brochure by Carmontelle, the author of all the society proverbs, entitled New Year's Conversations. What struck me as piquant in this work is its extreme lifelikeness. Nothing in society is as annoying as what we hear every day."[81]

It was with the same cynical tone that Carmontelle assumed the role of art critic, writing what has been called "the most prominent body of radical criticism produced before the Revolution."[82] In doing so, he followed his ailing friend Diderot, with whom he had been discussing art since the 1760s. Describing Lagrenée's four paintings representing the Virgin and the Child that hung at the Paris Salon in 1765, Diderot wrote: "Our friend Carmontel says that all this is pastiche, and he is wrong (etc)." But, at the end, Diderot avows: "How is it then that one of these children is beautiful and the other so mediocre? I could answer, but I won't. For Carmontel it would be easy."[83]

Using the dialogue form that he had mastered, Carmontelle analyzed the exhibited paintings. His reflections were relevant and pertinent, particularly with respect to composition and the use of perspective. He also, and perhaps above all, studied the public in attendance, linking the quality of the art to the merits of the society that produced it. As such, his salon critiques were another means of commenting on the world of the ancien régime. His anonymously published *Sideswipe at the Salon of 1779, dialogue, preceded and followed by thoughts on painting*, begins whimsically: "I stopped in the Tuileries to look at a fine statue that adorns the edge of the largest pool when I noticed a manuscript at my feet; curious to know what it contained, I opened it and saw: *Ideas on Painting*. I liked the title, it was a splendid morning, and so I sat down in the shade on a green lawn and read this stranger's thoughts."[84] But he shifts quickly to reflecting on his own fate as an artist: "As soon as the necessity of making a living takes his choice of means away from him, he dedicates all his abilities to a rich man; the rich man in his turn demands to be amused in his idleness, that he be entertained so as to forget the misfortune of being idle, requiring that his luxury or even his debauchery be abetted!"[85] His words are a biting echo of his proverb *Le Portrait*.

His pamphlet on the 1783 Salon recalled the opening lines of his *Garden at Monceau*: "I cannot abide these jealous minds who are saddened or wearied by praise of others, nor these beardless little experts for whom the faculty of thinking is a humiliating obligation, nor these listless automata who are insensible to the charms of painting because the beauties of nature have never touched them, nor even, if it needs saying, these presumptuous men of letters whose total ignorance of all the arts is considered an incontestable qualification for judging them well."[86] Five years earlier, hadn't he affirmed that "he who is the least knowledgeable is always the one who judges & decides with the greatest confidence" (*Garden at Monceau*, 14)? Educated and virtuous people are needed when it comes to admiring great art: "The fact that Raphael and Poussin brought the science of expression to its most eminent degree certainly proves the beauty of their genius, but it also reflects commendably on the men of their time." Regrettably, that was not the case in Paris under Louis XVI: "Unfortunately, the bad taste of the majority of wealthy people still resists universal approbation and makes it impossible for the author of a fine work to enjoy both true glory and material profit."[87]

Carmontelle's gaiety and humor are always present, however. He is astonished at the lack of inspiration in a history painting by Joseph-Marie Vien, an eminent member of the Académie royale: "Isn't it time that the lovely Helen, who has posed so many times, took some rest? Sitting, standing, reclining, from behind, from in front, in profile, in half profile, this Greek household has appeared in every aspect; to finish exhausting the combination of their attitudes, there remains only one way, which is to paint them with their feet up in the air."[88] But in 1789 he was enthusiastic about Jacques-Louis David's *Brutus*: "This is the first picture of genius to have come out of the modern school . . . I dare to maintain that if revolutions contribute to developing great sentiments in Artists, it is the great sentiments of the Artists themselves that prepared these revolutions."[89] The tempestuous heat of the summer of 1789 announces the storm, and Carmontelle admires the great artist who knows how to dominate his subject. "How beautiful it is to be able to hide in half-tones the most odious hero of a denatured patriotism."[90] If he sensed something to fear in his contemporary, he was right. Three years later David's bold signature would grace each page of the ominous register kept by the committee that sent the condemned to the guillotine.[91]

Inventing in the eye of the storm

A decade after publication of the brochure on Monceau, violence broke out in France, first in Paris and then at the frontiers of the realm. The king and queen were guillotined, and their cousin—Carmontelle's pupil, by then duc d'Orléans, his futile prince—followed them to the scaffold. The Orléans children were exiled, while the duchess took refuge in Normandy with her father, the duc de Penthièvre, who had remained very popular. The Orléans properties were seized, but the prince had already sold his collection of paintings in England. Carmontelle who had to leave the Palais-Royal, went to his sister Marguerite-Marie's at Goupillières, in the Chevreuse Valley, not far from Dampierre, where the duc de Luynes also had retreated to his château.

Once calm returned, Carmontelle went back to Paris and took lodgings in the rue Vivienne, near the Palais-Royal (which had become the Palais National), where he continued to work and invent. In 1795 he submitted his monograph on transparencies to the Committee of Public Safety, suggesting that the technology could be used for educational purposes, and also painted a long, transparent scroll depicting life throughout the four seasons of the year, a nod to the *Conversations* he had recently published.[92] Each scene shows busy workers reaping the summer's harvest, for example, or gathering wood for winter, while the gentry look on. In a reversal of roles, a château's windows are illuminated during a nighttime party to show the dancers within; the villagers look on in fascination in the darkness outside. It was thus that Carmontelle attempted to justify a revolution that had gotten out of hand.

He also sent the government a "Project de revêtement pour les murs des terrasses des Jardins des Tuileries" (Scheme for covering the walls of the terraces in the Tuileries Gardens) that involved engraving the names of all those who had died for the nation, along with their titles to glory, on giant marble slabs—a war memorial, in other words.[93] In addition, he sent the revolutionary assembly two manuscripts on perspective, which together constituted a treatise on the subject. The first, titled "La Perspective démontrée à l'usage des jeunes gens qui savent la géométrie et le dessin" (Perspective for young people who know geometry and design) contained thirty-seven pages and twenty fold-out plates (so that the images could be consulted next to the text). The second, "Extrait des éléments de géométrie applicables à la perspective démontrée" (Excerpt of the elements of geometry applicable to perspective), contained fifty pages and eight illustrations (fig. 18).[94]

Carmontelle's text appeared in the wake of works on perspective by Edme-Sebastien Jeaurat (*Treatise on Perspective for the use of Artists* of 1750) and Johann Heinrich Lambert (*Free Perspective*, published in 1759).[95] Wanting his manuscript to be accessible to young people, he used images to make studying more attractive: "The Convention has very wisely decreed that the study of geometry shall form part of

FIG. 18 Carmontelle's treatise on perspective, 1794–95.

the public education of young citizens because it is the base of all the sciences, arts, and trades. It serves in the formation of a healthy judgment, in encouraging the love of truth, in teaching the principles of nature as well as in studying all nature's products and in bringing about the enjoyment of all its charms."[96] He underscores the importance of geometry for architects and painters in particular: "The geometrician sees objects with precision as they really are; he turns them around; he penetrates them and calculates their dimensions. Painters see only appearances—that is to say, surfaces—for that is what they must solely represent. Yet, in order to make them exactly sensible to the gaze, they must be considered in the manner of a geometrician."[97] In stressing the practical applications of geometry, he addressed garden designers as well: "It could happen that while composing a scene [in a landscape], and having placed the main folly in the foreground, one finds oneself needing to place accessory follies at the rear, which one wishes to rise only to a certain height, and one wants to know what their size will be. This is to say that having placed their roofs in the composition, one wants to know where their bases would touch the ground. Nothing will be easier to determine."[98]

The treatise demonstrates both Carmontelle's technical competence and the desire to educate that was so particular to the Enlightenment. But just like his other two submissions of Year 3 of the Revolution, it elicited no response from the revolutionary assembly—France was at war with herself and on all her frontiers.

The end of a world

When Monceau was confiscated, the Arts Commission decided to leave everything in place. But very soon, having noted the deterioration of the site, its members asked the Commission of Public Works to police it and even suggested that its hay be harvested "in the English manner" so as to be useful to the republic.[99] The Commission of Public Works replied that it would stop the misuses at the "maison Monceau" and have the damage repaired. Then the assembly decreed that "the houses and gardens of Saint-Cloud,

Bellevue, Monceau, Le Raincy, Versailles, Bagatelle, Sceaux, Isle Adam, and Vanves shall be kept to serve the enjoyment of the people and form establishments that are useful to agriculture and the arts."[100] On 27 Prairial Year 4 (June 15, 1796), the Arts Commission removed the pictures and precious art objects that were still at Monceau: a Michelangelo, two Veroneses, Gobelins tapestries, a figure "in the Egyptian style," statues made of white and colored marbles, Houdon's African woman, etc.[101]

Because the Madeleine cemetery had become full due to the number of executions, the authorities requisitioned a plot of land at the extreme northeast corner of Monceau for a mass grave. Beginning in March 1794 the remains of almost fifteen hundred people who had lost their heads on the guillotine were buried there. Among the corpses thrown pell-mell into quicklime, their severed heads filling the gaps, were those of Madame Elisabeth, the king's sister; Guillaume-Chrétien de Lamoignon de Malesherbes, who had defended the king at his trial just as he had defended Encyclopedists and Protestants; and revolutionaries Georges Danton, Louis-Antoine de Saint Just, and Maximilien de Robespierre. When the executions finally ceased, the Monceau grave was filled in and forgotten. The garden, which was too costly to maintain, was rented out to an amusement-park operator, but the follies were deteriorating fast. To attract crowds, spectacular balloon experiments were organized at Monceau. André-Jacques Garnerin, a pupil of the physicist Jacques Charles, tested the first parachute there.

At the end of 1801, after order had been restored to France, First Consul Napoleon Bonaparte asked the second consul, Jean-Jacques-Régis Cambacérès, if he would like one of the national properties as a country home, Bonaparte having reserved Saint-Cloud for himself. Cambacérès chose Monceau, perhaps because he still had links to the Orléans family, for whom he had acted as legal counsel. He commissioned the architect Pierre-Nicolas Bénard, a nephew of the visionary architect Étienne-Louis Boullée, to replace the ruined main pavilion with a neoclassical building sporting a pedimented portico. Those follies that had become too damaged—the Turkish and Tartar Tents, the brass-ring Carousel made of painted sheet metal, and the French Pavilions made of wood clad in trompe-l'oeil marble plaques—were demolished, the garden was remodeled, and a nursery planted. Once work had sufficiently progressed to ensure safety, the estate was opened to the public. Did Carmontelle come?

The old artist, who had never married though he seems to have courted one of his models, was still painting transparencies just as precisely and faithfully as ever. In them can be seen the changes in French parks where vast grassy expanses and natural caves have replaced the bright colors of the ancien régime's follies, in the same manner that flowing, white, high-waisted dresses have replaced the ruched-silk panier frocks of the old order.[102] He still entertained his friends with transparency sessions, Mme de Genlis reported.[103] And then, the day after Christmas 1806, Louis Carrogis, known as Carmontelle, passed away. Born two years after Louis XIV's death, he died two years after Napoleon's coronation. At once a spectator and actor in the great theater of society, he had seen and observed a world that was destined to come to an end. With his pen and brushes, he had done much to record it for posterity.

Monceau: forever à la mode

After he became arch-chancellor of the new empire, Cambacérès purchased a mansion in the faubourg Saint-Germain and abandoned Monceau. Napoleon planned to turn it into a "Chinese" garden and later envisaged moving the menagerie of the Jardin des Plantes there, but the birth of his son, the King of Rome, changed his plans, and therefore Monceau became a place of promenade. Empress Josephine went there, it is said, to see the heir she had failed to conceive. The emperor considered giving the park to the imperial children, but the Chaillot hill, rising above the Seine opposite the École Militaire, seemed a more inspired setting for the son of a great leader.

After Waterloo the Bourbons returned to France, and the Orléans heirs recovered their father's properties. For Louis-Philippe and his sister Adélaïde, Monceau was a reminder not of their childhood promenades, especially since the main pavilion was no longer the same, but of their father's sad final years, which had hardly been the most brilliant. The new duc d'Orléans moved the small circular temple at Monceau to an island at the end of the park surrounding his Château de Neuilly where it still stands today overlooking the Seine (fig. 19). When he came to the throne in 1830, Louis-Philippe gave all his properties to his children—experience had taught him the speed and violence with which the political situation could change in France.

And indeed, he in turn was deposed in 1848; but although exiled like him, his nine children and their descendants still possessed Monceau, one half through their father's gift and the other through the inheritance of their aunt Adélaïde, who had never

FIG. 19 The circular temple from Monceau on the Île de la Jatte in Neuilly.

married. But the new government contested Louis-Philippe's gift, since they had confiscated his personal property, and his heirs could only lay claim to their aunt's half.

Napoleon III commissioned the prefect of the Seine, Georges-Eugène Haussmann, to turn his capital into a modern city. Unfortunately there was no money. The plain around Monceau was attracting interest, and since the price of land was rising steeply the Orléans family had already planned to develop the site. Haussmann struck a bargain with two bankers, the brothers Isaac and Émile Péreire: they could develop one half of the park if they turned the other into a public garden. The brothers didn't wait to be asked twice. The engineer-in-chief of the Paris Department of Promenades and Plantations, Jean-Charles Adolphe Alphand, created today's Parc Monceau in a space occupying just a fifth of the original estate, designing the park in a fan shape around a toll house built in the 1780s by architect Claude-Nicolas Ledoux just outside

FIG. 20 The Naumachia in the Parc Monceau.

FIG. 21 The Pyramid in the Parc Monceau.

the garden.[104] The Naumachia was restored (fig. 20), and the river, considerably shortened, was renovated; the pyramid (fig. 21) and a couple of tombs, displaced by the lavish mansions that had invaded the former wood, found refuge on one of the lawns that were dotted with exotic saplings. Alphand followed the same recipe he had used at the Buttes-Chaumont and Montsouris: lawns, a few specimen trees, a handful of monuments, a rock garden, a cave, a waterfall, etc. The results were referred to as "English parks" to please the emperor, who was a great admirer of London's green spaces. When baptizing the new boulevard adjacent to the former garden, it was remembered that Malesherbes's remains lay beneath it, and so the road was given his name.

A fashionable place of promenade

In 1861, some nine decades after its first inauguration, the new Monceau was unveiled by Napoleon III with pomp and festivities. Surrounded by the rustling crinolines of the ladies of the court, the sovereign strolled along the park's paths, admired Ledoux's rotunda (whose height had been slightly raised to accommodate an apartment for the caretaker), contemplated the carefully restored Naumachia, congratulated Alphand, and complimented the nurserymen. Monceau became the new place of fashionable promenade. Émile Zola described the milieu of the wealthy cosmopolitans who built mansions on the edges of the park: the banker-brothers Camondo from Constantinople, who collected eighteenth-century art and furniture; Henri Cernuschi, the financier of Italian origin who had a passion for Far Eastern art and whose costumed balls recalled the park's princely past; the Andrés, the Jamesons, the Rothschilds, the Gouïns, and the chocolate manufacturer Menier.[105]

Less costly than those giving directly onto the park were the houses that attracted artists keen to enjoy this haven of greenery as well as the favors of their clientele who lived in the new neighborhood. Albert Wolff described the quarter in 1886: "Friday is the day that most artists have chosen for their receptions. On that day the avenue de Villiers is curious to see: all of Parisian high society can be found there . . . each going to visit his painter; in every house in the neighborhood there is an artist's studio . . . A painter's status can be gauged by counting the number of carriages parked in front of his house."[106] Édouard Detaille, Ernest Meissonnier, Louis-Ernest Barrias, François Flameng, Jean-Jacques Henner, Pierre Puvis de Chavannes and, a little further away because he was just starting out, John Singer Sargent rubbed shoulders there with Sarah Bernhardt and Edmond Rostand. It was perhaps while contemplating the Parc Monceau's pool that Claude Debussy imagined *Pelléas et Mélisande*, and Charles Gounod and Gabriel Fauré frequented the park's winding, verdant pathways.

In his book on Monceau, Carmontelle postulated that we "need only Gardens where nature is present in its most agreeable forms; the charm one must feel on entering should be perpetuated, & renewed in every way possible, so as to awaken in the soul the desire to see it every day & to understand it" (*Garden at Monceau*, 16). It was just such a yearning that overcame James Tissot, who appropriated the gracious curve of the Naumachia colonnade reflected in the water as the background for several of his compositions.[107] Claude Monet spent hours at Monceau studying the light filtering through the greenery, groups of women under their parasols, and children's games.[108] And the young Marcel Proust, whose parents lived nearby, came to learn the workings of society as he watched darting little girls and young boys in sailor suits rolling their hoops under the gaze of nannies tasked with teaching them the language and morals of the English.

English morals? Like a posthumous protest on Carmontelle's part at the sight of these manicured lawns where meadow flowers no longer bloomed and glades where stiff marble monuments had replaced the bright stripes of the Turkish and Tartar tents, a giant figure loomed above the trees brandishing the flame of liberty (fig. 22). On the advice of the engineer Gustave Eiffel, the giant statue designed by Auguste Bartoldi—a gift from the French to the American people to mark the centenary of their liberation from British domination—was made at a foundry just outside the park gates.[109] On July 4, 1884, representing his nation the US ambassador came to take official possession of it to the sound of a brass band. Disassembled for her Atlantic crossing, *Liberty* would be put back together on an islet in the bay of New York. Carmontelle had warned: "The true Art is being able to keep the stroller present through the variety of features without which he will go off to find, in the open countryside, what is missing for him in this Garden—the image of freedom" (*Garden at Monceau*, 16).

FIG. 22 The Statue of Liberty, visible from the Parc Monceau, ca. 1884.

ESSAYS

FIG. 23 Paris and its immediate surroundings, from Jean Delagrive's *Carte topographique des environs et du plan de Paris* (1735).

HISTORY BY DESIGN:
THE AESTHETICS OF TRANSFORMATION IN CARMONTELLE'S JARDIN DE MONCEAU

David L. Hays

KNOWN PRIMARILY through the elegant folio published in 1779, Carmontelle's design for the Jardin de Monceau is usually approached as if conceived all at once on a blank slate. However, the concept was developed over at least half a decade and was determined by a complex process of site formation. The parcels of land assembled were variously residential, commercial, and agricultural. Yet Carmontelle combined new and existing elements to imply a fictional past for the garden, as if it had come about not only through recent invention—of which there was much evidence—but also through a deeper process of transformation in which nature played a determining role.

Transformation was of considerable interest in garden culture at that time, as traditional paradigms of design were being challenged by new ideals of nature. But Carmontelle's approach was unusual because it indexed those alternatives—traditional and natural—in a relational way, manifesting an aesthetic impulse not accounted for in his writings. Historicized through an aesthetic of transformation, his version of natural design was contingent and eclectic, and it was therefore at odds with the ideal of universal nature embraced by many of his contemporaries. Yet, for that very reason, Carmontelle's work at Monceau is an especially interesting precedent for landscape architecture today when the legacy of natural design has become a stumbling block and rethinking aesthetics has become essential to the development of more sustainable approaches.

To begin with, Carmontelle did not choose the site on which he worked. Instead, the project began in 1769—two years before Carmontelle's involvement—as a commission from his young patron, Louis-Philippe-Joseph d'Orléans, at that time duc de Chartres, to a speculating architect, Louis-Marie Colignon.[1] The site was about a mile northwest of the Paris city limit in a rural tract between the villages of Le Roule and Monceau (fig. 23). Since 1724, the King's Council (*Conseil du Roi*) had prohibited development in that and other parts of the *couronne*—the zone of market gardens and quarries surrounding the city—fearing erosion of supply capacities and consequent inflation.[2] However, in the wake of the Seven Years War (1757–63), the government reversed its policy and incentivized development in the hope of stimulating the regional economy. Liberalization of trade in goods, new securities for building loans and property-related investments, and favorable options for long-term leasing—the *bail emphytéotique* (a lease of variable duration, up to ninety-nine years), *bail à vie* (a life-term lease), and *vente à vie* (literally "sale for life," a life-term lease in which the benefits and liabilities of ownership, such as tax responsibilities, were transferred along with use of the property, but the title itself was not)—expanded opportunities for owners, developers, and renters alike. For architects such as Colignon, speculation could produce a substantial annual income.[3]

On July 26, 1769, Colignon purchased a 3-acre lot near the village of Monceau from a butcher named Pierre Aubert. Anticipating sensitivity to speculators, Colignon kept the price down by using a local stone-cutter, Jean Flagère, as a proxy.[4] Five months later, on December 19, 1769, he signed a contract with the duc de Chartres—then only twenty-two years old—for the development of a pleasure retreat on the site (fig. 24, parcel A in figure 25). Colignon had credibility not only as an *architecte du Roi*—one of the many responsible for royal residences—but also for having recently completed a sumptuous and much larger residential property close by (eventually known as the hôtel de La Vaupalière after the marquis who leased it). On the lot near Monceau, Colignon agreed to construct within three years a pavilion-style residence and a surrounding *bosquet*-style garden (one formed primarily through the geometrical arrangement of trees and trimmed hedges). The duc de Chartres, in turn, acquired life use of the property (through a *vente à vie*) in exchange for three thousand livres per year and a single, upfront payment of twelve thousand livres.[5] At the termination of the agreement (on the death of the duc de Chartres or if he stopped payments), the property

FIG. 24 Colignon's plan for a proposed pavilion and garden near Monceau, attached to his 1769 contract with Chartres. North is at the lower left.

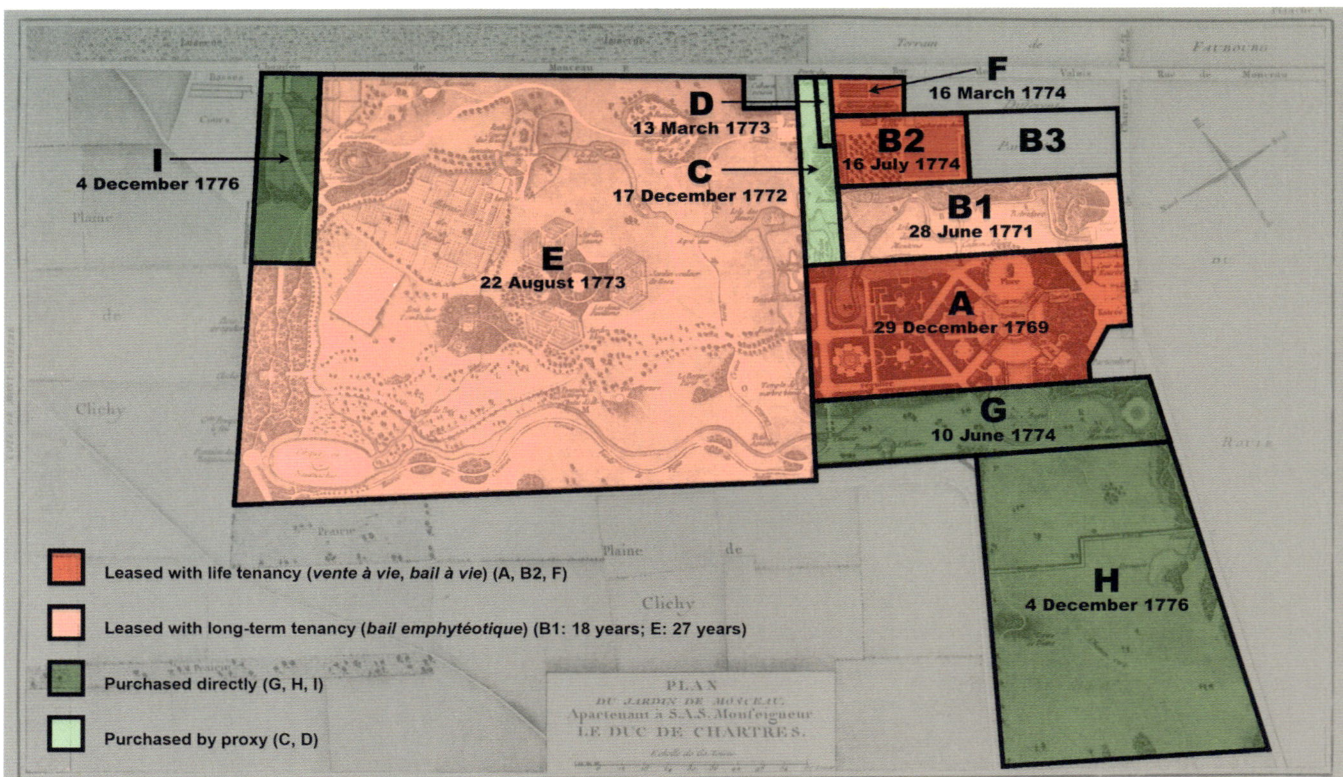

FIG. 25 Site formation at the Jardin de Monceau, 1769–76.

and all immovable improvements would return in full to Colignon or his heirs, an aspect of temporality that is usually overlooked in scholarship about Monceau.

Colignon completed the residence and garden within the agreed-upon period of time, but, halfway through that effort, on June 28, 1771, the duke expanded the site by leasing a 1.73-acre (0.7 hectare) lot along its southeast side (parcel B1 in fig. 25).[6] More specifically, a valet on his household staff, Jean-Baptiste Maublanc, and the valet's wife, Catherine Elizabeth Daler, acting as the duke's proxies, signed an eighteen-year sublease with the understanding—as in the original agreement between the duc de Chartres and Colignon—that any improvements to that site would eventually revert to its owners, Claude Brice Emery, *bourgeois de Paris*, and his wife, Anne Dessaudé.[7]

Instead of having Colignon expand the scope of his work, the duc de Chartres asked Carmontelle to transform the new lot into a garden. That choice revealed much about the kind of setting the duke had in mind. As a visual artist with training as an engineer (whether military or civil remains uncertain), Carmontelle understood landforms and how to depict topography.[8] Also, having served as general director of entertainments for the Orléans household since 1763, he had a reputation for inventiveness and resourcefulness in spatial arrangements.[9] In fact, Carmontelle may have organized one or more events at Monceau already, in the recently completed residential pavilion.

Carmontelle began work sometime after late June 1771, but in the following two years two more parcels were acquired (C and D in fig. 25). Lots in the *couronne* tended to be small-scale and highly diversified in both tenure and use, so the typical way to build a larger property there was through piecemeal accretion. In this case, both of the new parcels were purchased using Maublanc and Daler as proxies, but the smaller of the two lots, D—the 0.1-acre site of a drinking establishment (*cabaret*)—cost more than half as much as the much larger one, C (0.44 acre).[10]

Carmontelle completed his work sometime before July 12, 1773, when the condition of the site was detailed in a survey document (figs. 26 and 27).[11] As that drawing made clear, Carmontelle's approach contrasted sharply with that of Colignon. The portion of the *bosquet* closest to the entrance was replaced by a thicket, and an extended entrance path looped around to a circular arrival court framed by trees described elsewhere as horse chestnuts and Italian (Lombardy) poplars. In a corner close by, a large, spiral mount called the Belvedere provided views over the garden and its surroundings, while at the far end of the garden

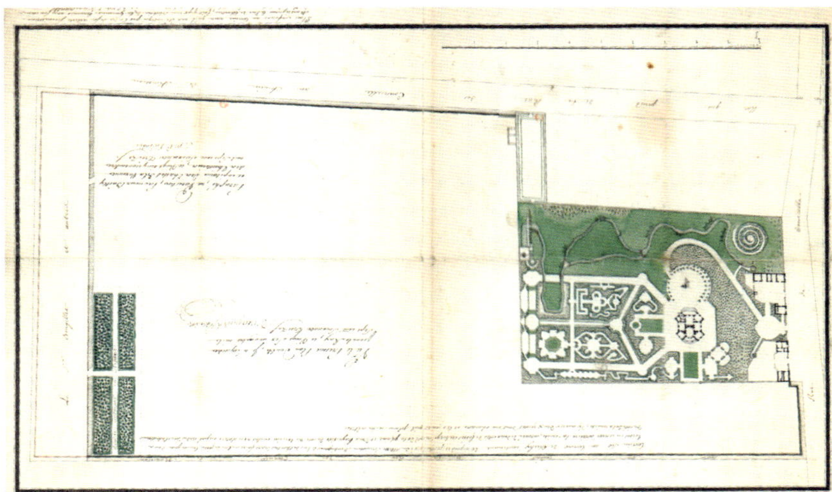

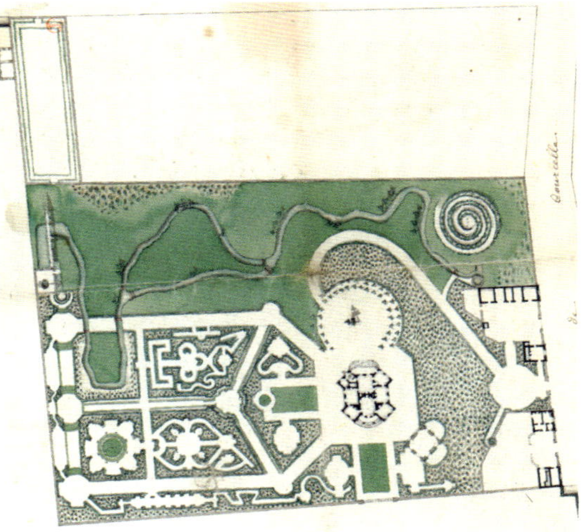

FIG. 26 Drawing, completed July 12, 1773, of the enclosure plan for the expanded Jardin de Monceau. It is rotated to correspond with the orientation of Carmontelle's 1779 plan of the garden, with north at the lower left.

FIG. 27 Detail of fig. 26, showing work completed by Colignon and Carmontelle.

Carmontelle installed a mock Water Mill and Bridge. The expanded garden was enclosed by walls on three sides, in some places obscured by thickets, but the boundary behind the Bridge was left exposed, evidently secured by a ditch, and the view there opened onto an extensive, cultivated field.

The surface of the expanded area was mostly lawn, and a stream meandered along its entire length, dividing in the northeast half of the plot to form a roughly triangular island. Near the northwest end, however, Carmontelle directed the stream into the *bosquet*, making it seem as if its natural course had overtaken, and thus transformed, one of the component spaces of Colignon's design. At the two places where the stream crossed—and thereby disrupted—an allée at the edge of Colignon's scheme, plank-style footbridges, set off-center, offered a seemingly improvised resolution, restoring the possibility of circulation there. Given that Carmontelle began work while Colignon was still completing the original commission, it remains uncertain if the part of the *bosquet* seemingly overtaken by the stream had been finished and was then transformed or if Carmontelle's intervention preempted that work. Also, the timing of the land acquisitions was such that the Water Mill and Bridge (both of which stood on parcel C) and the passage of the stream into the *bosquet* may have been conceived after an initial design (on parcel B1) was already underway or even realized. Whatever the circumstances, the stream historicized the garden as a whole by structuring an illusion of transformation in which nature played a determining role.

According to an entry for June 28, 1773, in the *Mémoires secrets*, the duc de Chartres's decision to create a garden at Monceau was motivated by envy over Tivoli, a garden belonging to the wealthy tax agent (*receveur général des finances*) Simon-Gabriel Boutin.[12] Situated along the rue de Clichy in Paris, in the vicinity of the present-day Gare Saint-Lazare, Tivoli was completed in 1771, around the time when Carmontelle began work at Monceau. Covering about 11.8 acres (4.8 hectares), it was divided into three distinctly styled sections separated by broad, tree-lined allées: a substantial vegetable garden covering nearly 4 acres (1.6 hectares), a *jardin italien* planted on a little over 1 acre (0.45 hectares), and a *jardin anglais* laid out over 3.7 acres (1.5 hectares). By mid-1773 the garden at Monceau did resemble Tivoli insofar as it featured contrasting regular and irregular components. The *bosquet* and new "natural" area at Monceau corresponded to the *jardin italien* and *jardin anglais* in the earlier garden. Like Carmontelle's later work at Monceau, the *jardin anglais* at Tivoli featured a stream and a roughly triangular island. Also, both of those settings epitomized the small scale, visual isolation, and component status (as parts of larger properties laid out in other, usually traditional styles) typical of early French experiments with natural or "English" garden design.[13] After that moment, however, the developments at Monceau took a very different course.

Soon after the completion of Carmontelle's work, the duc de Chartres began a campaign to enlarge the site even more. Over a three-year period, from August 1773 to December 1776, the area available for Carmontelle to design grew more than five-fold—from 5.27 acres to 28.42 acres (2.13 hectares to 11.50 hectares). Significantly, about 70 percent of that total was obtained not through purchase but through lease. Six separate parcels of land were added to the initial four, beginning with a large area to the northwest, which was the most consequential in size. Covering 14.20 acres (5.75 hectares) (see parcel E in fig. 25), this was the cultivated field seen beyond the Bridge in Carmontelle's initial work. It was owned by the nearby church of Saint-Philippe du Roule, which rented it to the duke through a twenty-seven-year lease (*bail emphytéotique*) (that is, to expire in 1800), likely to help finance construction of a new building.[14] That parcel also included a *remise* at its north end—a dense thicket planted to refuge game animals, a common topographical feature within princely hunting domains (fig. 26).[15] In fact, dozens of such spaces were scattered across the agricultural landscape near Monceau, and they were loathed by farmers because they sheltered game in proximity to crops (fig. 28). Accordingly, the decision to incorporate the *remise* into the garden, using ditches and fences to isolate it from its surroundings, would have been a welcome relief to local farmers.

On March 16, 1774, the duc de Chartres acquired another parcel from Emery and Dessaudé, a 0.19-acre (0.08-hectare) property obtained through a life-term lease at four hundred livres per year (see parcel F in fig. 25). Northernmost in a row of residential lots, it included a house and garden subsequently torn down and replaced with greenhouses. A few months later, on June 10, 1774, the duke purchased a 1.43-acre (0.58- hectare) lot on the other side of the original parcel—this time from Colignon himself, who had bought it earlier the same day for a significantly lower amount (see parcel G in fig. 25).[16] Colignon sold it to the duke, rather than leasing it to him, because of a clause in their original contract (that of December 19, 1769) specifying that any purchases of land adjoining the original lot would revert to Colignon or his heirs at the conclusion of the agreement—a condition evidently overlooked by the duke's administration and gravely disadvantageous to his estate. Recognition of that clause eventually led to costly renegotiations to protect the duke's investment at Monceau.[17]

The duc de Chartres had arranged a future lease of two additional parcels to the southeast (see parcels B2 and B3 in fig. 25). However, that deal was compromised when ownership changed hands. The duke was able to negotiate a life lease for one part (parcel B2, approximately 1 acre) with the new owners. That area abutted four lots he already controlled. But he never managed to lease the other part (parcel B3), which had street access and which the new owners retained for themselves.[18] Despite the duke's efforts, that area (parcel B3) was never included in the garden. Lastly, on December 4, 1776, the duke purchased two parcels at opposite ends of the garden from the same butcher, Pierre Aubert (by then retired), who had sold the original lot to Colignon in 1769 (see parcels H and I in fig. 25). Parcel H covered 4.65 acres (1.88 hectares) while parcel I included 1.68 acres (0.68 hectares).

Carried out between mid-1771 and late 1776, the gradual process of site formation had significant bearing on Carmontelle's eventual design (fig. 25 and plate I). For example, his initial work (parcels B1 and C), developed in relation to Colignon's bosquet garden, later became the entry sequence for the expanded garden, and what had first been an unexpected view over an adjacent field then became an opening onto the larger design (parcel E) (plate II). Around the corner, Carmontelle's setting for the Farm occupied the former site of the drinking establishment and the narrower, street-side portion of its neighbor to the north (Parcels C and D) (plate V). The two parcels behind the Farm (B2 and F), which were accessed through it, became kitchen gardens and greenhouses. Meanwhile, most of the thematic settings in the garden were distributed around the largest parcel (E)—an area acquired all at once—and seemingly conceived as a unity: a circuit loosely centered on a geometric garden (see fig. 25 and plates VI, VIII, and XII).[19]

As the site grew, Carmontelle repeatedly faced the challenge of animating spaces that lacked compelling natural features.[20] The areas to be incorporated were generally flat, and most had been used for agriculture, hence the appeal of adding thematic settings.[21] Still, Carmontelle did not treat the site as a blank slate. As noted, he integrated his initial work with Colignon's *bosquet*, cultivating an aesthetic of transformation. After the site expanded, he engaged with the *bosquet* yet again by reimagining it as a garden feature called the Regular Wood (*Bois Régulier*), a name noted on his plan of 1779 (plate I). The decision to retain Colignon's work was highly consequential, because it meant that the residential pavilion would remain visually isolated from most of the expanded garden.

Carmontelle also incorporated the abandoned *remise* at the far end of the garden, naming it the Irregular Wood (*Bois Irrégulier*), and he linked the Regular and Irregular Woods with a visual corridor, evident as a light line on the plan of 1779 (fig. 29 and plate I).[22] As a graphic element and an implied spatial condition, that line was defined not through a change in surface treatment (e.g., a gravel path across a lawn) but rather through careful framing and openness, even as footpaths crossed it multiple times and trees interrupted it in two places.

FIG. 28 Detail from Roussel's map of Paris and its surroundings (1730–39) showing *remises* in the vicinity of the village of Monceau.

Carmontelle's subtle, newly invented allée aligned with the long axis of Colignon's *bosquet*, which was centered on the northeast side of the residential pavilion. In Colignon's design, a small portion of thicket blocked the middle stretch of the axis, but Carmontelle's design cleared that area to open a line of sight to and from the building. At the opposite end, the new visual corridor led to a notch in the Irregular Wood, above which loomed a freestanding bell tower, fostering the illusion that a village was tucked in among the trees (plates IX and XI).[23]

Linked to the main pavilion, the visual axis extended by Carmontelle was figured as if it were a remnant of an earlier, traditional garden on the site. Of course, that was pure invention, as the area had recently been a cultivated field. But as a landscape version of a mock ruin, this "ghost allée" structured an impresssion of historical depth, as if the garden had come about in part through a process of transformation.[24] As in the first stage of Carmontelle's work, nature seemed to play a formative role in that shift, historicizing the garden as a whole, although the formal terms were here reversed. In the earlier instance, a natural line (the stream) disrupted a traditional condition (Colignon's bosquet), whereas in the later instance a traditional line (the ghost allée) persisted within a natural condition (see figs. 27 and 29).

Transformation was a significant practical concern in French garden design during the 1770s and 1780s as traditional priorities—including axial composition, bilateral symmetry, and the regularization of surfaces—were being challenged by new, "natural" ideals. For example, the anonymous *Lettre sur les jardins anglois*, published in 1775, asked readers to compare and assess two approaches for transforming, from traditional to natural style, a vast estate with a château by François Mansart and gardens by André Le Nôtre. The first of the two proposals relied on transitional spaces (*intervales, passages*), such as terraces with vases and statues and lawns in regular shapes, to integrate the formality of urban architecture with the "rural" character of natural gardens.[25] Among the devices listed were "aligned allées, that, losing their decorative aspect, through subtle gradation, terminate at rural thickets"—in the same manner as the ghost allée at Monceau.[26] The author's estate manager (*régisseur*) considered this "integrated" approach to be "more pleasant than a purely English one," not to mention "infinitely less destructive and less costly" and, through contrasts, more intriguing (*piquante*) and enjoyable.[27] Similarly, while learning about garden design in Paris during the 1770s and early 1780s, Francesco Bettini explored ways of converting regular layouts into irregular ones with a special interest in the reshaping of working gardens.[28]

But an *aesthetic* of transformation is different from the practical *accommodation* of transformation, because the aesthetic depends on the persistence of the past through traces, whether real or simulated. In other words, the aesthetic of transformation is *relational*, as in a palimpsest, and Carmontelle's way of figuring nature as a historicizing device therefore differed radically from the globalizing ideal of universal nature invoked in much contemporary theory and practice—for example, in Whately's *Observations on Modern Gardening* and at the marquis de Girardin's vast estate at Ermenonville. In that view, nature was transnational and ahistorical, so to design in imitation of nature was to transcend traditional styles.[29] In contrast, Carmontelle's approach to natural design was contingent and eclectic. One way to see that difference is by comparing his de-

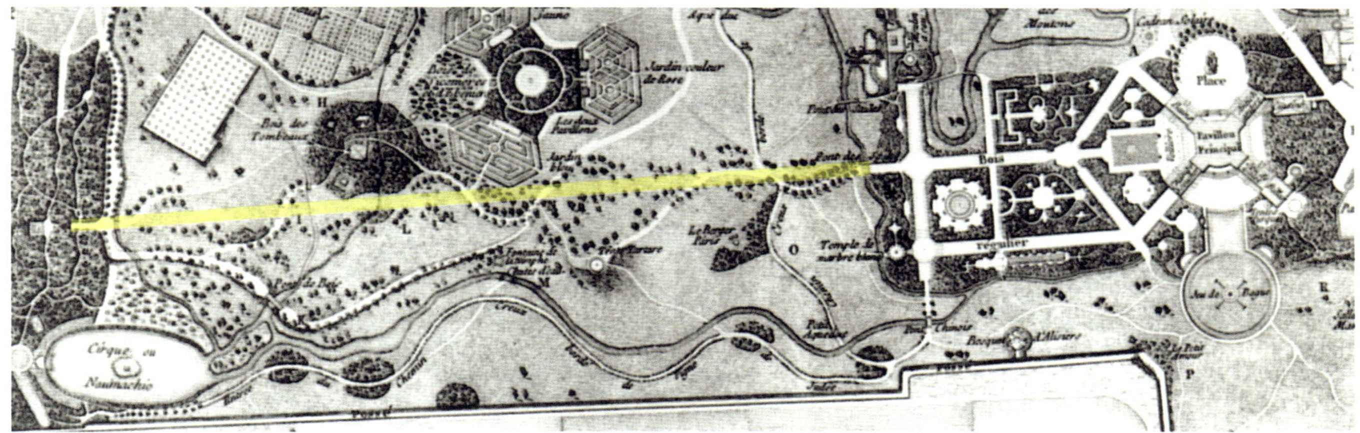

FIG. 29 Detail from Carmontelle's plan of the Jardin de Monceau showing the visual corridor linking the Regular and Irregular Woods.

sign for Monceau with an anonymous plan of the garden as it was likely transformed soon after its completion (fig. 30). In 1783 the Scottish plantsman Thomas Blaikie assumed direction of planting and design at Mon-

ceau, and he began reorganizing the garden in keeping with his sense of nature as a generalized and generalizing condition.[30] In the anonymous plan, the *bosquet* (Regular Wood), the *remise* (Irregular Wood), and the ghost allée have been eliminated. Other regular landscape features, such as the geometric garden, the circular platform for the Chinese Merry-Go-Round (brass-ring carousel), and the circular arrival court are also gone, and the shape of the Circus or Naumachia has been adapted to make it seem more like a natural pond. In other words, "historical" traces, embodied in geometrical forms, were systematically erased.

Thinking about landscape has changed significantly since the late-eighteenth century, yet landscape architecture is still haunted by expectations that it be mistaken for nature. To design with nature now means to embrace continual change, but the aesthetic terms appropriate to that practice are still being defined. Given that context, Carmontelle's way of figuring transformation presents an interesting example for landscape architecture today—a precedent for "palimpsestic" works, such as Gas Works Park (Seattle, Washington), the Promenade Plantée (Paris, France), and their manifold progeny, including Duisburg-Nord Landscape Park (Duisburg, Germany) and the hugely popular High Line (New York, New York), in which cultural artifacts (e.g., remnant infrastructure) and "natural" plantings (e.g., those that seem spontaneous) are closely correlated.

FIG. 30 Anonymous plan of "a garden at Monceau," ca. 1788. Here the image is rotated to correspond with the orientation of Carmontelle's plan of the Jardin de Monceau, with north at the upper left.

FIG. 31 *Madame la duchesse de Chartres* in 1770, by Carmontelle.

FASHION FOLLIES AT THE *FOLIE DE CHARTRES*

Caroline Weber

THE PLATES for Carmontelle's *Jardin de Monceau* reveal not one but two major advances in eighteenth-century French design. The first and best known of these of course has to do with the innovative landscaping carried out under the patronage of Louis-Philippe-Joseph d'Orléans, duc de Chartres, and aimed, as Carmontelle writes in his explanatory text from 1779, at evoking *"all times and all places."* The attire worn by the female figures in his engravings—an eclectic array of styles whose names referenced far-off regions like Poland and Asia—would appear to conform to this ideal. Yet they reflect an aesthetic development that occurred quite independently of Carmontelle, albeit contemporaneously with his work on the Monceau project. Characterized by Daniel Roche as a "revolution in fashion," this development involved a radical change in Frenchwomen's dress, away from the ossified formality of court attire and toward newness and whimsy.[1] Louise-Marie-Adélaïde de Bourbon-Penthièvre, duchesse de Chartres (1753–1821), helped to catalyze this shift through her patronage of a forward-thinking designer named Rose Bertin (1747–1813). By representing styles akin to those Bertin devised for the duchesse de Chartres and her friends and imitators, Carmontelle subtly documents a second *folie de Chartres*: the sartorial follies brought into vogue by his patron's wife.

When the duc de Chartres and mademoiselle de Penthièvre wedded at Versailles on April 5, 1769, neither party was marrying for love. As the daughter of the duc de Penthièvre, a grandson of Louis XIV through one of that king's bastard sons, the bride carried a stain of illegitimacy that her union with a prince of the blood—Chartres was a scion of the royal Orléans clan—promised to mitigate. The groom was drawn to the young lady's wealth, as he was a notorious spendthrift and she the heiress to the largest fortune in France. Her dowry included a staggering sum of six million livres and an income of nearly a quarter-million livres a year.[2]

For all its expediency, the match was tainted by scandal. Rumor held that a year earlier the duke had poisoned one of his closest friends, the infamously decadent prince de Lamballe—mademoiselle de Penthièvre's only sibling—so as to concentrate even more wealth in his future consort's hands.[3] In fact Lamballe died of syphilis, but his sister did inherit his estate, and Chartres spent liberally from the funds she brought to the union. No sooner had they married than he began buying up the tracts of land around the village of Monceau that would eventually form his vast private garden. In this practical sense, the Jardin de Monceau would have been unthinkable without the duchesse de Chartres.

Beyond this, however, it is unlikely that the duchess had much direct influence over the project. Their marriage was *désuni* ("dis-united")—a favorite French term for ill-starred conjugal pairings—from the start. Her riches may have induced Chartres to marry her, but they also enabled him to carry on the life of wild dissolution he had enjoyed as a bachelor.[4] The duke was so unabashed in what his kinsman Louis XV euphemistically called his "bad habits" that the king tried to discourage the betrothal, warning the father of the bride: "[Chartres] is a libertine. Your daughter will be unhappy."[5] Given the monarch's own well-earned reputation for depravity, this was a chilling forecast.

It turned out to be true. Despite the considerable pressure he was under to father an heir to his title and estates, Chartres all but ignored his new wife, expending his prodigious erotic energies instead on a rotating cast of paramours who ranged from courtesans and actresses to the wives and daughters of his fellow grandees.[6] The flagrancy of his philandering would do much to foster the contemporary perception of Carmontelle's Monceau as a grand-scale *garçonnière*—a fantastical venue for its owner's salacious exploits.[7]

As for the duchess, her husband's infidelity made it difficult for her to fulfill her own putative mission as a bearer of princes: though she gave birth to the couple's first child in the summer of 1772, the infant was a stillborn daughter and thus doubly unfit to perpetuate the Orléans line. Left to her own devices, she befriended three young women at court whose domestic circumstances were as unhappy and unfruitful as hers. The members of the coterie were her widowed sister-in-law, the ethereal princesse de Lamballe, whose womanizing husband may have given her syphilis, if not a child, before his untimely death from the disease; the duchesse de Lauzun, a delicate beauty similarly ill-used by her adulterous spouse (another of Chartres's accomplices in debauch)[8]; and Marie-Antoinette, whose husband, the future Louis XVI, likewise shunned the marriage bed, if out of timidity and anxiety rather than extramarital distraction.[9] This group of friends coalesced shortly after Marie-Antoinette's arrival at court in May 1770, when the dauphine was fourteen, the duchesse de Chartres seventeen, and mesdames de Lamballe and de Lauzun in their early twenties.

Together these neglected young wives turned to fashion as their prime distraction and raison d'être. One of their fellow courtiers described Marie-Antoinette in terms that applied as well to the other members of the quartet: "To be the most à la mode woman alive seemed to her the most desirable thing imaginable."[10] And it was the duchesse de Chartres who not only sparked this desire in her friends but presented them with an irresistible way to satisfy it. For her wedding trousseau she had relied on the services of Bertin, a then obscure *marchande de modes* working in a tony boutique in Paris. So taken was she with Bertin's inventiveness and flair that the duchess promptly claimed her as her protégée and introduced her to her high-ranking companions, who also became devoted customers.[11]

As a *marchande de modes,* Bertin belonged to a relatively new class of French garment workers whose labors complemented but were distinct from those of seamstresses and tailors. Predominantly female, they formed an offshoot of the mercers' guild, specializing in "readily modifiable, ever-changing trimmings and ornaments designed to alter and enhance already made dresses and skirts."[12] As Clare Haru Crowston has shown, the *marchandes de modes* emerged as a powerful market force in the last quarter of the eighteenth century in response to a newfound demand for "elements of personal taste, choice, and superfluity" from female clothing consumers.[13] Before then, elegant dress had changed little from year to year; in large part it was slow to evolve because it was so expensive to make. The *robe à la française*, the compulsory *habit de cour* for the female denizens of Versailles, was the ultimate case in point. This ornate, elaborately constructed garment comprised a long train, wide panniers, and a stiff, embellished stomacher worn over whalebone stays. As the costume was far too costly to be jettisoned after a season's wear, the ladies of the court tended to keep their *robes à la française* for many years, even decades, on end. The duchesse de Chartres donned a rose-colored habit of this kind when sitting for a portrait by Carmontelle in 1770, but she and her friends chafed at the fustiness of the style and hankered for an alternative (fig. 31). By designing accessories and adornments that made tired old dresses look new, Bertin and her competitors gratified this craving and established novelty and change as the motors of modern fashion.

While she belonged to the same generation as her ducal and royal patrons—Bertin was twenty-one to the duchess's sixteen at the time of the Chartres's wedding—their support catapulted the young marchande de modes to the top of her profession. By October 1773 she had grown successful enough to open her own boutique in Paris, the Grand Mogol. Located in the rue de Saint-Honoré—then as now one of the city's most fashionable districts—the shop quickly became a beacon of la mode parisienne. Demand for Bertin's creations reached even more dizzying heights when Marie-Antoinette became queen in 1774 and effectively anointed her as her "minister of fashion."

Bertin's clients were drawn to her innovations in two areas: headdresses and traditional "French" dress silhouettes.[14] Examples of her work in both categories abound in Carmontelle's renderings of the *Jardin de Monceau*. For instance, the center female figure in the view of the White Marble Temple (plate XV), the far-right woman in the view of the Farm (plate V), both women in the view of the Italian Vineyard (plate IX), and all four women in the view of the Wooden Bridge (plate X) wear versions of Bertin's first major millinery invention: a jaunty headdress known as the *quesaco*, which drew its name (the Provençal expression for "what's up?") from the title of a satirical work by Pierre

FIG. 32 Fashion plate showing a back view of the *quesaco* headdress, 1778.

Caron de Beaumarchais. The *quesaco* was made up of one or more long, artfully curled plumes worn at the back of the head, positioned between two vertical rows of sausage-roll curls and above a beribboned chignon (fig. 32).[15] This novelty item sparked an unprecedented craze among the women of Versailles; thus coiffed they courted attention in Paris by strolling along bustling thoroughfares and through popular city parks. While the *quesaco* trend predated Monceau's completion by a good half a decade, that setting was just the sort of high-profile arena in which Bertin's devotees liked to parade their fabulous modes.

In the spring of 1774 Bertin launched an even more ostentatious headdress: the pouf. It would spawn countless variants in the years that followed, but in all cases the pouf was built on a gigantic scaffolding of teased hair, horsehair pads, and chicken wire. Measuring up to three feet high and held together by multiple coats of pomade, this avatar of the beehive served as a pedestal for a seemingly infinite variety of decorative flourishes. The first recorded sighting of this creation was of the *pouf au sentiment*, a homage in hair to events or persons of special importance to the wearer, and it appeared on the head of the duchesse de Chartres. In April 1774 she fêted the recent and long-overdue birth of her first son by going to the Paris Opéra in a pouf that displayed a woman seated in an armchair holding an infant at her breast: "this [was] the duc de Valois and his wet-nurse. On the right, a parrot, a precious bird much prized by [the duchesse, was] picking at a cherry. On the left [was] a little Negro resembling her beloved African page," Almanzor.[16] This coiffure gave the duchess one awkward moment when its gauze wrappings snagged on a girandole, knocking the parrot and cherry out of position.[17] Yet the pouf's obvious unwieldiness did not sour her or her would-be emulators on the look. Quite to the contrary, wrote one observer: "All the ladies went mad for this pouf and wanted to have one."[18]

Without delay, the companions of the duchesse de Chartres set out to surpass her in all things pouf.[19] On May 10, 1774, Louis XV died of smallpox, leaving Louis XVI and Marie-Antoinette to ascend the throne. Unlike in the new queen's native Austria, the smallpox vaccine was viewed with suspicion in France. Braving a fierce popular outcry, Marie-Antoinette prevailed upon her husband to undergo the vaccination. To celebrate her victory, she appeared in a *pouf à l'inoculation* featuring the serpent of Aesculapius, the Greek god of medicine, coiled around an olive tree that symbolized wisdom (whether that of modern medicine or its royal champion was anyone's guess).[20] Not to be outdone, the duchesse de Lauzun commissioned a pouf that depicted "a stormy sea, ducks swimming along the shore, a huntsman waiting for the right moment to shoot them, a windmill where the miller's wife was being seduced by a village priest, and, at the bottom, the miller walking along with his donkey."[21]

Setting aside the wish-fulfillment aspect of its tableau of spousal infidelity (making the husband, not the wife, the betrayed party), Mme de Lauzun's pouf bears noting because the windmill that stood at its apex was also one of the Jardin de Monceau's signature follies, sited on the property's highest point. It is fitting, then, that in his view of the Dutch Windmill as seen from the Isle of Rocks (plate VII), Carmontelle should have included two pouf-topped ladies, although like the creators of the era's leading fashion chronicle, the *Gallerie des modes et costumes français,* he chose to portray somewhat less intricate versions than those favored by Bertin's most august patrons. The woman with the parasol on the far left of the Dutch Windmill engraving wears a *pouf à l'Asiatique*, so named by Bertin to appeal to the contemporary French taste for chinoiserie and *turquerie*. (This same taste, not incidentally, accounts for the "Chinese" and "Turkish" garb of the male domestics in Carmontelle's Monceau engravings—as well as in the duc de Chartres's actual service.) The high silk crown of this headdress was adorned with feathers, tassels, bows, and flowers. Sometimes looping strands of pearls were added, though none are visible in the Carmontelle image. This "Asiatically" coiffed figure strolls alongside a woman wearing a less overwrought but no less voguish *bonnet au pouf,* an oversized white mobcap festooned with ribbons and trailing ruffled lappets. The view of the Temple of Mars (plate VI) shows another accessory of this type, worn by the central female figure (figs. 33 and 34).

The other Bertin creations that pervade Carmontelle's Monceau renderings are a cluster of dress types based on her sensational *robe à la polonaise*, its triple-swagged overskirt a saucy allusion to the partitioning of Poland by Prussia, Austria, and Russia.[22] The puffed swags of this garment, held in place by tassels or bows, revealed a large swath of the underskirt in back; a cutaway of the overskirt produced a similar effect in front (fig. 35). While retaining the whalebone stays and panniers also required by the robe à la française, the polonaise and its variants—the *circassienne* (outfitted with shorter swags and a more dramatic cutaway than the "Polish" model) and the *turque* or *musulmane* (distinguished by a trailing overskirt, an underskirt of contrasting fabric, and long, as opposed to three-quarter-length, sleeves)—were hailed as welcome correctives to that dress because they did away with the cumbersome train and shortened the hem of the underskirt to bare the feet and ankles. As a result, they were considered exceptionally well-suited to outdoor promenades—whether in the parks of Paris or in the gardens of the Petit Trianon (the king's gift to Marie-Antoinette upon their accession and a favorite gathering spot for her and her

FIG. 33 Fashion plate showing a *robe à la circassienne* and a *pouf au fichu*, ca. 1785.

inner circle).[23] The *Gallerie des modes* duly equipped its *po-lonaise*-, *circassienne*-, and *turque*-clad beauties with para-sols and walking sticks, symbols of their newfound free-dom of movement (figs. 36 and 37).

Here again, Carmontelle's Monceau would function as an ideal backdrop for the new fashions, as his engrav-ings make clear. Whereas the sixteen images that repre-sent female figures include only eight *robes à la française*, they contain fourteen *polonaises*, twelve *circassiennes*, and four *turques*.[24] And even the more conventionally garbed women sport samples of the millinery Bertin made fash-ionable. For instance, the Marie-Antoinette look-alike who faces the viewer in the image of the Ruined Castle (plate IV) has accessorized her formal court gown with a fanciful pouf designed just for her, its cluster of plumes dyed black—one of the Habsburg monarchy's heraldic colors—in tribute to her family of origin. (One cannot tell in the engraving if these plumes are ostrich feathers: if they are, then they effect a visual pun that intensifies the

FIG. 34 Fashion plate showing a *bonnet au pouf*, 1776.

"Austrian" dimension of the coif; *autruche* ["ostrich"] is a near homonym of *Autriche*.) As her husband's reign progressed, touches like these would take on increasingly provocative political significance for Ma-rie-Antoinette, her friends, and her enemies alike. But when Carmontelle drew up his plans for Monceau, that aspect of the "revolution in fashion" still lay in an unforeseen future. For the time being, and in his style-conscious renderings, it was only figuratively that the queen and company had lost their heads.

FIG. 35 A French *robe à la polonaise*, depicted ca.1780.

FIG. 36 Fashion plate showing a taffeta *robe à la circassienne*, 1779.

FIG. 37 Fashion plate showing a *robe à la polonaise*, 1778.

FIG. 38　Josef Kohaut (Kohault) in 1764, by Carmontelle.

CARMONTELLE'S PORTRAITS OF MUSICIANS

Florence Gétreau

Scholars studying Carmontelle in recent years have concentrated in particular on his "garden of illusions"—a subject magnificently pioneered by Laurence Chatel de Brancion. Yet the literature devoted to the garden—its history, plantings, sculptures, and constructions—neglects one element. What did its visitors hear? How did they enjoy the garden for their musical entertainment?

Carmontelle himself supplies us with clues. His portraits of musicians—some sixty in all—help us to envision the soundscape of the Jardin de Monceau (and those of other Parisian gardens). A good half of the musicians he portrayed are depicted near an arcade opening onto a park; approximately ten are placed on a terrace giving onto trees; and another ten or so are found at the edge or in the middle of gardens or, far less often, in forests. Only three are shown in interiors. Four of these musicians were composers, three were professional performers shown in isolation, eleven were professionals portrayed in groups of various sizes, and the remainder—the vast majority—were amateurs.[1]

Carmontelle's portraits have received relatively little attention since Gruyer's seminal 1899 volume showcasing those in the Chantilly collection.[2] Nor has there been any systematic investigation of the subset of portraits featuring musicians; only a few have received more than minimal commentary. In 1929 Georg Kinsky, in the French version of his *Album musical*, reproduced both a caricature engraved by Carmontelle in 1760 of Rameau walking in the garden of the Palais-Royal and Delafosse's famous 1764 print of the Mozart family.[3] François Lesure, in his pioneering work on the social history of music as seen through images, was the first to address the subject of aristocratic "connoisseurs" in his comment on these two portrait sketches.[4] Far more developed was Albert Pomme de Mirimonde's analysis of twenty-eight Carmontelle portraits with respect to the instruments being played and their popularity among the aristocracy in his 1977 survey volume on French musical iconography.[5] In 1988 I gave a short introduction to Carmontelle's portraits of musicians in relation to five of his group portraits in the Musée Carnavalet that were shown in my exhibition *Instrumentistes et luthiers parisiens*. I divided images of composers, professional performers, and the private sphere of high society into separate groups to help reveal the vogue for certain instruments.[6] Finally, much more recently (2017), Mary Cyr published a methodologically progressive article about three portraits of musicians: one showing Mlle Pitoin and her father; another depicting the demoiselles Royer, daughters of the composer Pancrace Royer; and the third a portrait of the marquis d'Ussé.[7]

Carmontelle's portraits of composers: from reflection to inspiration

Carmontelle portrayed several composers. Jean-Philippe Rameau is shown towards the end of his life working in an armchair on the edge of a terrace, near a harpsichord whose shape and decoration recall those of his contemporary, the Paris-based German instrument maker Jean-Henry Hemsch.[8] Relaxed but deep in concentration, the composer is caught in the act of writing a manuscript which, undecipherable to the beholder, could be either music or theory. The harpsichord is closed, its cover strewn with books and sheet music. In the foreground, carelessly lying on the floor, is Rameau's theoretical work *Code de musique pratique, ou Méthodes pour apprendre la musique*, which he published in 1760, the year the drawing was made, and whose title, or at any rate the beginning of it, is easily decipherable. While intellectual reflection characterizes this portrait, another, etched the same year by Carmontelle, focuses not without humor on the composer's lanky silhouette and expressive concentration. In this little caricature, unique in Carmontelle's oeuvre, Rameau is shown walking in the garden of the Palais-Royal, bent over and deep in thought, completely oblivious to the trivial straw chair nearby that is included to evoke the public space of promenade. Baron Grimm, in his obituary of Rameau published just after the composer's death in 1764, said of him: "He was as remarkable for his figure as famous for his work. Much taller than M. de Voltaire, he was just as pale and emaciated. Since he was constantly to be seen out walking in public, M. de Carmontelle drew him from memory a few years ago. This witty little engraving is a very good likeness."[9] The portrait would go on to be very successful, being variously copied and decked out with new accessories (trees, a young girl on the chair, an arcaded house). It was even used on the cover of a manual for debutant musicians, *Le Petit Rameau,* which of course was not the work of the great man himself.

After Rameau, Carmontelle drew the portrait of Egidio Duni, shown seeking inspiration at the harpsichord (the same black model with gilded trimming), his absent gaze lifted towards the light. He was the in-house composer at the Théâtre-Italien from 1760 to 1770; it seems probable that his likeness dates from this period, when he composed so many successful comic operas to French-language libretti, before being overtaken by Pierre-Alexandre Monsigny, François-André Danican Philidor, and André-Ernest-Modeste Grétry. Carmontelle drew Monsigny standing, in a dove gray coat with golden yellow braiding, with nothing to indicate his profession. Behind him is a bridge in the sloping garden of a large house with a dovecote.

FIG. 39 Mme de Maupassant in 1759, by Carmontelle.

Like Carmontelle, Monsigny had a connection with the Orléans court. On first arriving in Paris, when he was *Receveur général du Clergé de France* and training as a musician with Pietro Giannotti (d. 1765), he entered into contact with the duc d'Orléans (senior), who, a few years after Monsigny's first successes at the Théâtre-Italien (*Le Roi et le Fermier,* 1762; *Rose et Colas,* 1764), made him his *maître d'hôtel.* It is perhaps this new role that Carmontelle's portrait was intended to celebrate just before the enormous success of Monsigny's comic opera *Le Déserteur* (1769), which premièred at the hôtel de Bourgogne.

Another composer Carmontelle depicted was one of the last virtuoso lute players of Europe, Josef Kohaut (or Kohault), who was born in Vienna and moved to Paris in 1762 as a musician in ordinary to Louis-François de Bourbon, prince de Conti.[10] He is shown sitting comfortably in an armchair in a shady park (fig. 38), his crossed legs supporting a large lute with thirteen courses. It was no doubt made in Germany, given its stringing, chanterelle pulley, and three sound holes, which recall the work of Prague lute maker Thomas Edlinger, who built models inspired by the early-seventeenth-century Venetian theorbos of Magno Tieffenbrucker, updated for new repertoires.[11]

A young Viennese, Count Wenzel Anton von Kaunitz-Rietberg, ambassador to France in 1750–52, probably played a crucial role in introducing Kohaut and his elder brother, Karl, also a lute player, to Paris. Grimm wrote about the two ten years later: "This Mr. *Kohaut* [i.e., Josef] has an elder brother who came to France with Count von Kaunitz and who is sublime when he plays the lute. The Kohaut brother who stayed here also plays this same instrument, but coldly and unenthusiastically—the man of genius is in Vienna."[12] In the service of Prince Hieronim Florian Radzivill in Biala Podlaska and Slutsk (today in Poland and Belarus, respectively) from 1753 to 1759, Josef afterwards spent six months in Italy. As of July 1762 he settled in France, in the service of the prince de Conti. He began performing at the Concert Spirituel (sixteen times from 1763), where he played duos for lute and cello as well as motets of his own composition.[13] Between 1764 and 68 he wrote three works for the Comédie Italienne and contributed to two others.

Professional performers

Carmontelle drew three famous Parisian opera singers, of whom one was Sophie Arnould. He depicted her in the role of Thisbé in the fifth act of *Pyrame et Thisbé*, a tragic opera by Rebel and Francoeur. Premiered in 1726, *Pyrame et Thisbé* was revived in 1759, with more than forty performances up to 1761, and again in 1771, with eighteen performances in February and March. In the final act, when Thisbé sings "Amour! que ton flambeau me guide," the libretto stipulates that "the Theater represents a thick Wood: we see, through the trees, the Tombs of ancient Assyrian Kings."[14] Carmontelle's drawing bears the (apocryphal?) date 1760, which suggests that the stage set represented may have been that used for the first revival. Chatel de Brancion, however, suggests that what we see here is perhaps the second revival of 1771, since she sees a link between the background and Carmontelle's drawing for the twelfth plate of his *Jardin de Monceau* (1778), *View of the Wood of Tombs*, in which one of the graves in question, as in *Pyrame and Thisbé*, takes the form of a pyramid. Either way, Carmontelle beautifully captures Arnould's moving performance and "the astonishing expression of her acting" (as praised by the *Mercure de France* in February 1759). The newspaper's reviewer noted that her role "gave to the voice and soul of the young Actress every possibility of touching and charming the Spectators."

Mlle Chevalier, the second opera singer Carmontelle depicted, was drawn at the end of her career in two theatrical postures that adroitly suggest the expressive roles of witch and Fury that were often given to her. As for Regina Mingotti, Carmontelle met this Neapolitan singer, who worked in Dresden, Munich, London, and at the Spanish court, during one of her short stays in Paris. She performed at the Concert Spirituel, singing a program of Italian airs for the feast of Corpus Christi in 1752; took the stage again in September 1754; and performed in Pergolese's *Stabat Mater* for Palm Sunday 1760—the year Carmontelle depicted her. The image shows her standing by a harpsichord in a far more familiar manner than in Raphaël Mengs's 1745 portrait of her, now in Dresden.

Also in 1760 Carmontelle portrayed the three daughters of the late composer and harpsichordist Joseph-Nicolas-Pancrace Royer, who directed the Concert Spirituel from 1748 to 1754. At the edge of a terrace, near a very tall circular topiary arcade, the three young women are playing together. On the music stand is an open score on which one can read the title *Zaïde*, a highly successful three-act ballet written by their father. Is this allusion just a posthumous homage? Indeed, none of the daughters seems to be singing, and their late father does not seem to have published a selection of airs from his ballet. This has led the musicologist Mary Cyr to query the combination of instruments visible in the portrait—violin, harpsichord, and guitar—which, she points out, were used by Pierre de Lagarde, one of Royer's colleagues at the Musique du Roi, in his 1764 *cantatille* (i.e., vocal piece) *Amarillis*.[15] She also points out that the rondeau *La Zaïde*, transcribed in Royer's *Pièces de clavecin*, offers a fairly simple structure in three parts that could have been played by the trio shown in this image.[16]

Among Carmontelle's rare group portraits of professional musicians it is worth mentioning two from 1764. First is the very famous *Mozart Family*: Leopold on the violin, Marie Anna singing, and Wolfgang Amadeus at the harpsichord. There are four known versions of the portrait as well as an engraving that Jean-Baptiste Delafosse published the same year (see fig. 7). The print ensured continent-wide circulation for this spectacular image showing "Marianne Mozart, Virtuoso aged eleven and J.G. Wolfgang Mozart, Composer and Maître de Musique aged seven" under their father's command.[17] The other group drawing of musicians portrays the harpist Jean-Aimé Vernier (standing) reading sheet music, Jean-Pierre Duport (or his younger brother Jean-Louis) on the cello, Jean-Joseph Rodolphe on the horn, Pierre Vachon on the violin, and Ignace Prover on the oboe.[18] Not only is this image the first French visual source for the chamber role given to the orchestral horn (with the technique of putting the hand in the bell to ensure the full chromatic range),[19] it is also the only visual record of this famous ensemble of musicians who were in the employ of the prince de Conti.[20] The drawing should probably be dated to after 1769, when Jean-Louis Duport entered the prince's service.

Lady harpsichordists or Les Amusements du Parnasse (Michel Corrette, 1749)

In the study I undertook with Denis Herlin of portraits of eighteenth-century Parisian harpsichordists, we noticed a preponderance of excellent women players in the images, literary sources, and dedications of works for this instrument.[21] Some of these women have passed into posterity because they gave their names to harpsichord scores that preserve the memory of their virtuosity. One example is Mlle Pitoin (accompanied by her father on the viola da gamba), who posed for Carmontelle with the score of Jacques

FIG. 40 Mme de Montainville in 1758, by Carmontelle.

Duphly's *Médée* open on her harpsichord. In a tribute to her talents, Duphly composed a musical portrait of her. Titled *La Pothoüin*, it was written for the harpsichord with viol accompaniment ad libitum, the same combination of instruments depicted in this drawing.[22] Other harpsichordists are shown playing in highly stereotyped attitudes, their instruments more distinctive than their poses. Thus Mme de Maupassant is shown playing a ravishing harpsichord, its case "red inside with a sky-blue band on a white background" (using the terms employed in the sales advertisements of the time) (fig. 39). The support, with its cabriole legs, has sculpted decoration highlighted in the same blue, and all the colors match the player's floral dress perfectly. Mlle Desgots's instrument is "red inside" also, but has "gilded bands," while the exterior features garlands of roses on a gilt background. The most common model depicted by Carmontelle features bronze-finished or gilded bands on a black background. Sometimes the instrument is shown closed but with a musician nearby; in others it is reduced to an accompanying piece of furniture whose role is to symbolize the idea of sociability through music.

L'art de chanter (Jean Blanchet, 1756)

There are six portraits by Carmontelle of women who seem to be singers or musicians, because they have musical scores open on a stand or balanced on their knees. Of them, Mme d'Epinay is the most famous (Jean-Jacques Rousseau mentions her musical talents in his *Confessions* and she took lessons in composition with Charles-Louis Dupin de Francueil), while Mlle d'Avenart is considered a "virtuoso of the greatest might, playing at the Concerts de la Reine"—an affirmation that so far has proved impossible to confirm.[23]

Les Dons d'Apollon … avec l'histoire allégorique de la guitarre (Michel Corrette, 1763)

Guitar playing seems to have been a frequent female pastime in the entourage of the duc d'Orléans—between 1758 and 1784, Carmontelle depicted eight elegant noble ladies doing just that. Each is seated and has a guitar held in place by a strap attached to the headstock with a pretty red or blue bow; it features five double strings, a deep parchment sound-hole insert, and a *pistagne* pattern of marquetry in alternating ebony and ivory around the soundboard. All except one of the instruments depicted display the proportions of the models made by the Voboam dynasty of Parisian guitar builders, five members of which succeeded each other between 1630 and 1730. Their instruments are always recorded in revolutionary seizures of property.[24] Some of the guitars drawn by Carmontelle even feature the decorative ebony "mustache" they employed on either side of the bridge (fig. 40). A new model appears in the 1784 portrait of Mme Alexandre de Damas, displaying a more scooped-out body and a distinctly trapezoidal headstock that is tilted slightly backwards, corresponding perfectly to those Voboam guitars that have come down to us from that decade.

Guitars weren't the only plucked string instruments that were appreciated in Carmontelle's day. Firstly there is the mandora: a round-backed lute with either a curved or perpendicular headstock strung with five, six, or eight courses (according to an eighteenth-century German definition).[25] It comes as no surprise to find the baronne d'Holbach playing it, her instrument held in place by a blue strap. Mlle de Croimare is also playing a mandora; hers is strung with five courses and features three sound holes, and its body with alternating ribs recalls old Venetian lutes (like Kohaut's modern lute mentioned earlier). A little theorbo (or cittern-theorbo?) held by M. Le Dran, a small five-course cittern played by Mme de Meaux, and Abbé Allegri's Milanese mandolin demonstrate the diversity of plucked string instruments that attracted amateur players in the last third of the eighteenth century. Easy to learn, they were perfect for accompanying fashionable ariettas or the loveliest arias from comic operas.

L'Art de jouer de la harpe (Cardon, 1784)

The chromatic harp with tuning pins and pedals was perfected in southern Germany sometime around 1730. It came to France mid-century and was played at the Concert Spirituel by Georges-Adam Goepfert in May 1749 and in 1760 by Christian Hochbrücker (a Bavarian musician and instrument builder in the employ of the prince de Rohan) and Philippe Jacques Mayer. It is hardly surprising that the instrument was rapidly adopted by amateur musicians, ladies of the court, and the aristocracy. La Live de Jully had himself portrayed at the harp by Greuze in 1759 and by Carmontelle a year later, and Étienne Aubry painted Mme Victoire playing the instrument in 1773. Even more sophisticated models became all the rage at the court of Marie-Antoinette.

L'École d'Orphée
(Corrette, *Méthode de violon*, 1779)

Violins are fairly rare in Carmontelle's portraits. There is one almost carelessly cast onto the floor in the portrait of the marquis d'Ussé, along with sheet music bearing the names of Vivaldi and Tartini. Cyr has

FIG. 41 M. and Mme Blizet with the actor Le Roy, ca. 1765, by Carmontelle.

rightly pointed out the old-fashioned tastes of this amateur musician, who is shown gazing at scores of Lully's *Armide* and André-Cardinal Destouches' *Issé* that are open on his music stand. Significant—perhaps unprecedented—in Carmontelle's oeuvre is a pair of drawings depicting an unidentified standing woman playing the instrument, one showing her in three-quarter view and the other from the back. In the former, visible on the music stand is a score of a sonata by Tartini.

At a time when virtuoso Italian and French violinists dominated the concert circuit, amateur musicians often fell back on a hybrid instrument, the quinton, whose career marked a short period of transition. It enjoyed its hour of favor particularly among ladies, since it is easier to play than the violin and more becoming, because the arms do not need to be raised (it is supported on the knee). Like the pardessus de viole, the quinton has five strings—whence its name—that span a chord; frets on the neck, which are a useful guide for playing in tune; and a carved headstock. It resembles the violin in the form of its body, with its squarish shoulders, F-holes, slightly curved back, and curly bracket contours. Carmontelle depicted Mlle de Bernay playing such an instrument in 1764 with the bowing technique of the viol. She is sitting in a garden in front of a row of orange trees in tubs. Another drawing, dated 1758, shows Mlle Grimperel tuning up her quinton on her knees; the frets are clearly visible on the neck. Carmontelle is the only artist except Anne Vallayer-Coster (most notably with a still life in 1770 that was her *morceau de réception*) to have depicted this instrument at least three times.

La Vielleuse habile (François Boüin, 1761)
and Les Amusements champêtres (Nicolas Chédeville, 1729)

Initially the hurdy-gurdy had a trapezoidal body. In the 1730s it changed shape when urban lute makers recycled lutes and guitars to satisfy noblemen and ladies in search of bucolic amusement. The refashioned instrument comprised two melody strings shortened by tangents operated by a push keyboard and drone strings rubbed continuously by a wheel turned by a hand crank. Beautifully made, the hurdy-gurdy replicated the aesthetic of baroque guitars.

The instrument became immensely popular in chamber formations with a violin, tenor instruments, and harpsichord accompaniment (as corroborated by the abundance of teaching manuals and sheet music). Carmontelle contributed to this aristocratic triumph by depicting Mme de Serré demonstrating her skill to Mme de Julienne in a drawing dated 1760.[26] The hurdy-gurdy is often associated with another drone instrument (i.e., instrument capable of producing continuous sustained notes), the musette with bellows. In this very sophisticated version of the bagpipe, air is introduced with a bellows held under the arm rather than by blowing into the instrument. Multiple drones are condensed into a very convenient cylindrical shuttle drone, while the chanters are equipped with keys that allow for chromaticism and ornaments. Charpentier

(116)

played a hurdy-gurdy, accompanied by Danguy on the musette, at the Concert Spirituel in 1732 and 1733.

The musette's precious materials (ebony and ivory) made it a masterpiece of refined craftsmanship, and its fabric-covered bag could be matched to the player's outfit. It was suitable for pieces performed with harpsichord accompaniment, or given *concertante* with the violin, flute, or hurdy-gurdy. After Watteau painted his faux village minstrels, many indeed were the aristocrats who had themselves portrayed as Arcadian shepherds playing the musette (for example, the marquis de Gueydan, as painted by Rigaud). Carmontelle took a different approach, depicting the instrument in the hands of Mme Blizet, who is accompanied by her husband on the violin (fig. 41), while the actor, Le Roy sits attentively. Apart from *Le Colin-maillard*, painted in 1736 by Jean-Baptiste Pater (Berlin, Charlottenburg), this is a rare case of an image showing the musette played by a woman.

Hunting instruments were played chiefly by men as well. Carmontelle had the opportunity to observe enthusiasts of hunting to hounds utilize a type of horn known as the *trompe Dauphine*.[27] He portrayed the instrument in the hands of highly experienced horn blowers, including the duc d'Orléans himself (see fig. 4) and his master of the hunt, M. de Bois-Massot. Both show us how to carry the instrument without hindrance, whether on foot or on horseback; huntsman M. de Champignelles seems ready to put his to his lips.

Carmontelle's most unexpected portrait of a musician must surely be that of Narcisse, the duchesse de Chartres's black servant (fig. 42). Wearing a sword and a black coat with the arms of the duc d'Orléans on the pocket, he is blowing into a galoubet—a pipe with three holes that is played with just one hand—all the while banging his snare *tabor* (much like a *tambourin de Provence*). The latter is not shown in the image, but it was indispensable in forming the inseparable rhythmic and melodic duo of dance instruments used at public and aristocratic balls and on the stage at the opera since the era of André Campra.[28] The large brown shadow cast onto the stone balustrade illusionistically evokes its throbbing presence.

Carmontelle depicted certain famous musical personages without their favorite attribute: including the vicomtesse de Beaumont and Mme de Lamballe, who were talented harpists;[29] Princess Pälffy-Kinsky, whose harpsichords and pianoforte were mentioned in the revolutionary inventories of her rue St.-Dominique mansion;[30] and Benjamin Franklin, promoter of the bewitching glass harmonica and inventor of the lightning rod, is shown as the inspiration behind the constitution of Pennsylvania (see fig. 8). Also, many instruments are absent in Carmontelle's portraits of musicians: chamber organ, clarinet, bassoon, and serpent; trumpet, trombone, and double bass; timpani and military drums; triangle and cymbals. These lacunae reflect the fact that music was exclusively an amusement among high society; Carmontelle displayed an undeniable bias towards the favorite instruments of the aristocracy. As his portraits suggest, the extent to which music and musicians contributed to the activities in the Orléans family residences cannot be doubted. To the magic of the garden scenes he drew, a fourth dimension should be added— the sonorous music that filled them.

FIG. 42 Narcisse, a servant of the duc d'Orléans, by Carmontelle.

✢

Carmontelle's Portraits of Composers, Musicians and Other Individuals, 1758–1785, Grouped by Musical Attributes

	PERSON *Notable composers, singers, and musicians*	IMAGE TYPE	DATE	LOCATION	INSTRUMENT(S), *accessories, sheet music or music books, other attributes*
1	Jean-Philippe Rameau (composer)	drawing	1760	Chantilly, Musée Condé (hereafter Chantilly), Car415	Harpsichord, score; *Code de musique pratique*
2	Idem	engraved caricature	1760	Paris, Bibliothèque nationale de France, Estampes AD-18-FOL and N2	Walking, without musical attribute
3a	Leopold, Wolfgang, and Marie Anna Mozart	drawing	1764	London, British Museum, 1994.0514.48	Harpsichord (Wolfgang), violin (Leopold), score (Marie-Anne)
3b	Idem	drawing	1764	York, Castle Howard	Idem
3c	Idem	drawing	1764	Paris, Musée Carnavalet, D4496	Idem
3d	Idem	drawing	1764	Chantilly, Car418	Idem
4	Leopold and Wolfgang Mozart	drawing	1766?	Location unknown	Harpsichord, violin
5	Egidio Romualdo Duni (composer)	drawing	n.d.	Chantilly, Car416	Harpsichord
6	Regina Valentini Mingotti (opera singer)	drawing	1760	Chantilly, Car431	Harpsichord, score
7	[Royer] The three daughters of composer Joseph-Nicolas-Pancrace Royer (musicians)	drawing	1760	Paris, Musée Carnavalet, D4507	Harpsichord, violin, guitar, score (players unidentified)
8	Josef Wenzel Thomas Kohaut (composer and lutist)	drawing	1764	Chantilly, Car426	Lute
9	Jean-Pierre or Jean-Louis Duport, Jean-Joseph Rodolphe, Pierre Vachon, Ignace Prover, and Jean-Aimé Vernier (musicians)	drawing	after 1769	Chantilly, Car424	Cello (Duport), horn (Rodolphe), violin (Vachon), oboe (Prover), score (Vernier)
10	Monsigny, Pierre-Alexandre (composer)	drawing	n.d., after 1768?	Chantilly, Car417	Standing (no attribute)
11	Sophie Arnould (opera singer)	drawing	1760	Chantilly, Car420	Shown in the role of Thisbe in *Pyramus and Thisbe* of Francoeur and Rebel
12	Mlle Marie-Jeanne Fesch Chevalier (opera singer)	drawing	n.d.	Chantilly, Car422	Dressed as Hebe?
13	Idem	drawing	n.d.	Chantilly, Car423	Holding a cane

	PERSON *Professionals and amateurs*				
14	Mme the duchesse de Bourbon, Mmes de Barbantane, de Hunolstein, de Vauban, and de Fitz-James	drawing	ca. 1760–1765	Paris, Drouot, June 24, 1981.	Harpsichord (player unidentified)
15	Mlle Desgots de Saint-Domingue and her black servant, Laurent	drawing	1766	Paris, Musée Carnavalet, D.4498	Harpsichord, score
16	M. and Mme Despourdons	drawing	1766	Paris, Musée Carnavalet, D.4499	Harpsichord (M.), score
17	Mme de Maupassant	drawing	1759	Chantilly, Car294	Harpsichord, score
18	M. Philippe, Mlle Delon, and M. Tellier	drawing	ca. 1763	Paris, Musée Carnavalet, D.4500	Harpsichord (Delon), violin (Philippe), score
19a	Mlle Pitoin and her father	drawing	after 1768	Chantilly, Car432	Harpsichord (Mlle), viola da gamba (father), score of Jacques Duphly's *Médée*
19b	Idem	drawing	n.d.	Paris, Sotheby's, Sep. 14, 2017. no. 38	Idem
20a	Mme the marquise de Rumain, with her daughters the comtesse de Polignac and Mlle de Rumain	drawing	n.d. (ca.1768)	Chantilly, Car322	Harpsichord (Mlle), score
20b	Idem	drawing	n.d. (ca. 1768)	London, Colnaghi, May 2000	Idem
21	Mother and her daughters (unidentified)	drawing	n.d. (1785–88?)	Chantilly, Car478	Pianoforte (daughter), score
22a	M. de La Live [de Jully] (master of ceremonies, art collector)	drawing	1760	Chantilly, Car349	Pedal harp
22b	Idem	drawing	ca. 1760	Montecarlo, Sotheby's, 1989, private collection	Pedal harp
23	M. and Mme Longueil	drawing	1769	Chantilly, Car79	Pedal harp (attribute present; not being played)
24	Mme Moreau, Mlle de Flinville	drawing	1762	Chantilly, Car304	Pedal harp (Moreau), music stand, score
25	Mme la comtesse de Polignac	drawing	1771	London, Sotheby's, June 30–July 1, 1988, no. 83	Pedal harp
26	Mlle Privée	drawing	n.d.	Chantilly, Car433	Pedal harp
27	M. le chevalier de Dreneuc	drawing	1771	Chantilly, Car65	Violin
28	M. and Mme Lallemant de Nantouillet, their son, and MM. de Damas and Lallemont de Lévignen	drawing	1772	Chantilly, Car440	Violin (Lallemant de Nantouillet)
29	M. and Mme Blizet with Monsieur Le Roy (actor)	drawing	ca. 1765	Washington, DC, National Gallery of Art, 1987.36.1	Violin (M.), musette (Mme)
30	M. the marquis d'Ussé	drawing	1760	Chantilly, Car198	Violin, music stand, music books

31a	Woman (unidentified) seen from front	drawing	ca. 1758–60	New York, Metropolitan Museum of Art, 2019.138.2	Violin, double music stand, score
31b	Same as above, seen from rear	drawing	ca. 1758–60	New York, Metropolitan Museum of Art, 2019.138.2	Violin, double music stand, score
32	Mme la comtesse d'Auxy	drawing	n.d.	Chantilly, Car209	Guitar and music book
33	Marie Josephine Catherine Collet, comtesse Alexandre de Damas-Tredieu	drawing	1784	Chantilly, Car238	Guitar
34	Sophie Septimanie de Richelieu, the comtesse d'Egmont	drawing	1758	Chantilly, Car253	Guitar
35	Mme de Montainville	drawing	1758	Amsterdam, Rijksmuseum, RT-1961-18	Guitar
36a	Mme Rigaud de Vaudreuil	drawing	1759	Chantilly, Car335	Guitar
36b	Idem	drawing	n.d.	Paris, Galliera, June 21, 1963; Paris, Drouot, March 11, 1988, no. 40	Idem
37	Mme the vicomtesse de Rochechouart	drawing	1760	Chantilly, Car319	Guitar
38	Mme de Villaumont and Mme d'Escours	drawing	n.d.	Chantilly, Car341	Guitar (player unidentified)
39	Two young women (unidentified)	drawing	n.d.	Paris, Musée Carnavalet, D.4497	Guitar (player unidentified)
40	Woman, with a child, playing a guitar (unidentified)	drawing	n.d.	Paris, Sotheby's, September 14, 2017, no. 77	Guitar, music stand
41	M. l'Abbé Allegri (diplomat)	drawing	n.d.	Chantilly, Car358	Lombardic or Milanese mandoline
42	M. and Mlle de Croimare and an unidentified young man	drawing	1766	Chantilly, Car143	Mandora (Mlle), violin (young man), double music stand
43	Mme the baronne d'Holbach	drawing	1766	Chantilly, Car403	Mandora, music stand, score
44	Mlle de Bernay	drawing	1764	Chantilly, Car214	Quinton
45	Mlle de Grimperel	drawing	1758	London, Christie's, January 13, 1993, no. 92	Quinton
46	Young woman in profile, seated in a garden (unidentified)	drawing	n.d.	Paris, Sotheby's, September 14, 2017, no. 17	Quinton
47	Mme and Mlle de Meaux and M. de Saint-Quentin (artist)	drawing	ca. 1758	Chantilly, Car429	Cittern (Mme), flute (Saint-Quentin), music book (Mlle)
48	M. Le Dran	drawing	n.d.	Chantilly, Car428	Cittern-theorbo or theorbo (?)
49	Mme de Julienne, Mme de Serré	drawing	1760	Paris, Musée Carnavalet, D.4506	Hurdy-gurdy (Serré)
50	Narcisse	drawing	n.d.	Chantilly, Car85	Pipe and tambourine of Provence
51	M. le duc d"Orléans	drawing	1763	Chantilly, Car2	Hunting horn
52	M. de Bois-Massot	drawing	1764	Chantilly, Car44	Hunting horn
53	M. de Champignelles (bailiff)	drawing	n.d.	Chantilly, Car54	Hunting horn
54	Mlle d'Avenart (opera singer)	drawing	n.d. (not after 1762)	Chantilly, Car421	Desktop music stand with score

	PERSON				Attribute associated with sitter, but not in portrait
55	Mme d'Epinay	drawing	1759	Chantilly, Car256	Music stand, score
56	Mme the princesse royale de Hesse-Darmstadt	drawing	1769	Chantilly, Car19	Score
57	Mlle de La Hulière	drawing	1759	Chantilly Car278	Score
58	Mme Le Fèvre	drawing	1780–85?	Chantilly, Car287	Italianate music book
59	Mme the baronne de Talleyrand	drawing	n.d.	Chantilly, Car326	Music stand, score
	PERSON *Notable musicians represented without attributes*				*Attribute associated with sitter, but not in portrait*
60	Mme the princesse Kinsky	drawing	1765	Chantilly, Car272	Harpsichord
61	Mme la vicomtesse de Beaumont, called in error Mlle du Beaumont du Repaire	drawing	1759	Chantilly, Car212	Harp
62	Mme the princesse de Lamballe	drawing	n.d.	Chantilly, Car10	Harp
63a	Benjamin Franklin	drawing	ca. 1780-81	Washington, DC, National Portrait Gallery, Smithsonian Institution, NPG.82.108	Glass harmonica
63b	Idem	engraving	n.d.	Blérancourt, Musée franco-américain du Château de Blérancourt	Idem

Index of Musicians Drawn by Carmontelle and their Musical Attribute

Musical attributes associated with the sitter in Camontelle's portraits are noted in parentheses. When multiple sitters are portrayed, but the identification of sitter to attribute is not established, the attribute depicted in the portrait is italicized. Musical instruments that an individual is known to have played, but that are not illustrated, are enclosed in brackets.

Numbers following names refer to Carmontelle's portraits in Appendix 1.

Composers

Duni, Egidio Romualdo (1708–75) (harpsichord), 5

Kohaut (or Kohault), Josef Wenzel Thomas (1734–77) (lute), 8

Monsigny, Pierre-Alexandre (1729–1817) [violin], 10

Mozart, Leopold (1719–87) (violin), 3, 4

Mozart, Wolfgang (1756–91) (harpsichord), 3, 4

Rameau, Jean-Philippe (1683–1784) (harpsichord), 1, 2

Rodolphe, Jean-Joseph (1730–1812) (horn), 9

Professional musicians and performers

Arnould, Sophie (1740–1802) (opera singer), 11

Chevalier, Mlle. See Fesch, Marie-Jeanne

Duport, Jean-Pierre (1741–1818) or Jean-Louis (1749–1819) (cello), 9

Fesch, Marie-Jeanne, known as Mlle Chevalier (1722–after 1789) (opera singer), 12, 13

Mozart, Marie-Anne (Nannerl) (1751–1829) (score), 3

Mingotti, Regina Valentini (1722–1808) (opera singer) (harpsichord, score), 6

Prover, Ignace (1727–74) (oboe), 9

[Royer], the three daughters of Jean-Nicolas-Pancrace Royer (c. 1705–55) (*harpsichord, violin, guitar, score*), 7

Vachon, Pierre (1738–1803) (violin), 9

Vernier, Jean-Aimé (1769–after 1838) (score, [harp, violin]), 9

Amateur musicians and connoisseurs

Allegri, M. l'abbé (diplomat) (Lombardic or Milanese mandolin), 41

Auxy, Marie-Louise de Monceaux, marquise de (guitar, music book), 32

Avenart, Mlle de (virtuosa at the Concerts de la Reine) (singing, desktop music stand, score), 54

Barbantane, marquise de, and her two daughters (*harpsichord*), 14

Beaumont, Mme la vicomtesse de [harp], 61

Bernay, Mlle de (quinton), 44

Blizet, M. (violin) et Mme de (musette), 29

Bois-Massot (Boismassot), Guillaume-Marin de Rouil de (head of the hunt for the duc d'Orléans) (hunting horn), 52

Bourbon, Louise-Marie-Thérèse-Bathilde d'Orléans, duchesse de (1750–1822) (*harpsichord*), 14

Champignelles, M. de (bailiff) (hunting horn), 53

Collet, Marie-Joséphine-Catherine. See Damas-Trédieu, Alexandre de

Croimare, Louis-Eugène, marquis de (no attribute), 42

Croimare, Elisabethe-Thérèse, mademoiselle de (lute or mandora, double music stand, score), 42

Damas-Trédieu, Marie-Joséphine-Catherine Collet, comtesse Alexandre de (guitar), 33

Damas, M. de (relation of Lallemant, see below) (no attribute), 28

Delon, Mlle (de Genève) (harpsichord), 18

Desgots, Mlle, de Saint Domingue (harpsichord), 15

Despourdons, M. et Mme (harpsichord in background with open score), 16

Dreneuc, chevalier de (violin), 27

Egmont, Sophie Septimanie de Richelieu, comtesse d' (guitar), 34

Epinay, Louise-Florence-Pétronille-Tardieu d'Esclavelles, Mme La Live d' (1726–83) (music stand, score), 55

Escours, Mme d' (*guitar*), 38

Fitz-James, Marie-Claudine-Sylvie de Thiard de Bissy, duchesse de (*harpsichord*), 14

Flinville, Mlle de (no attribute), 24

Franklin, Benjamin (1706–90) [glass harmonica], 63

Grimperel, Mlle de (quinton), 45

Hesse-Darmstadt, Caroline, princesse royale de (score), 56

Holbach, Charlotte-Suzanne d'Aine, baronne d' (1733–1814) (mandora, music stand, score), 43

Hunolstein, Charlotte-Gabrielle-Elisabeth-Aglaé de Puget de Barbantane, comtesse d' (*harpsichord*), 14

Julienne, Mme de (no attribute), 49

Kinsky, Marie-Léopoldine-Monique, comtesse Pälffy d'Erdöd, princesse (1729–94) [harpsichord], 60

La Hulière, Mlle de (score), 57

La Live de Jully, Ange-Laurent (1725–79) (master of ceremonies, art collector) (pedal harp), 22

Lallemant de Lévignen, Charles-Louis-François (relation of following) (no attribute), 28

Lallemant de Nantouillet, Charles-Marie-François Xavier, M. et Marie-Adélaïde de Damas-Crux, Mme (no attributes), 28

Lallemant de Nantouillet, Alexandre-Marie-Louis-Charles (son of previous) (violin), 28

Lamballe, Marie-Thérèse-Louise de Savoie-Carignan, princesse de (1749–92) [harp], 62

Le Dran, Henry-François (1685–1770) (surgeon) (theorbo or cittern-theorbo), 48

Le Fèvre, Mme (music book), 58

Le Roy, M. (actor), 29

Longueil, M. and Mme (pedal harp), 23

Maupassant, Mme de (wife of a military officer) (harpsichord), 17

Meaux, Mme (cittern) and Mlle de (music book), 47

Montainville, Mme de (*guitar*), 35

Moreau, Mme (pedal harp, music stand, score), 24

Narcisse (black servant boy of the duchesse de Chartres) (pipe and tambourine of Provence), 50

Orléans, Louis-Philippe, duc d' (1725–85) (hunting horn), 51

Philippe, M. (violin), 18

Pitoin (or Pothoüin), M. (viola da gamba) et Mlle (harpsichord), 19

Polignac, Gabrielle-Yolande-Claude-Martine de Polastron, comtesse, then duchesse de (1749–93) (pedal harp), 25

Polignac, Constance-Gabrielle-Bonne de Rumain de (eldest daughter of the marquise de Rumain) (no attribute), 20

Privée, Mlle (pedal harp), 26

Rigaud de Vaudreuil, Mme (guitar), 36

Rochechouart, vicomtesse de (guitar), 37

Rumain, marquise de (no attribute), 20

Rumain, Mlle de (harpsichord, score), 20

Saint-Quentin, M. de (artist) (flute), 47

Serré, Mme de (hurdy-gurdy), 49

Talleyrand, baronne de (music stand, score), 59

Tellier, M. (no attribute), 18

Ussé, Louis-Sébastien Bernin, marquis d' (violin, music stand, music books), 30

Vauban, Henriette de Puget de Barbantane, comtesse de (*harpsichord*), 14

Villaumont, Mme de (*guitar*), 38

Unidentified individuals

A mother and her daughters (pianoforte, score), 21

Two young women (*guitar*), 39

Woman playing a violin (violin, double music stand, score), 31

Woman, with a child, playing a guitar (guitar, music stand), 40

Young woman in profile seated in a garden (quinton), 46

Young man (violin), 42

FIG. 43 Carmontelle presenting the keys of the Jardin de Monceau to the duc de Chartres, ca. 1778.

"THE HABIT OF SEEING THE SAME THINGS OFTEN:" PLANTING THE PICTURESQUE IN EIGHTEENTH-CENTURY FRANCE

Elizabeth Hyde

In 1779 Louis Carrogis de Carmontelle published his *Jardin de Monceau*, a text with engraved plates showcasing the garden he had designed for Louis-Philippe-Joseph, duc de Chartres. By the time the volume appeared, the garden—begun in the early 1770s—was a fully realized French expression of *a* French version (not *the* French version) of the Picturesque garden. Carmontelle's creation was, as David L. Hays has articulated using the words of Carmontelle himself, "not a *Jardin Anglais*."[1] It was not even a "French" interpretation of the *jardin anglais*—the sort of garden typified by that of René-Louis de Girardin at Ermenonville (1766–76). Those gardens, created through artful *subtraction* of excess woods and other elements that concealed vistas and careful accentuation of natural features to reveal the picturesqueness naturally inherent in the site (the "genius of the place") were quite the opposite (in theory and effect, at least) of the Monceau project. Here was a garden of *addition*: "In the Garden at Monceau, everything had to be created," Carmontelle wrote (prospectus, 8). The garden was constructed from plots of land transformed through the creation of numerous features designed to entertain.[2] Carmontelle wrote that residents of Paris ceased out of familiarity to be impressed by such extraordinary sites as the Louvre and the gardens of the Tuileries: "The habit of seeing the same things often makes us ignore their value (*Garden at Monceau*, 14)." His task, then, was to ignite curiosity; to delight and surprise through the creative additions to the landscape. He wrote, "Since everything must be created, let us use this freedom to enchant, to amuse, & to arouse interest (*Garden at Monceau*, 14)." Such efforts might backfire. "People therefore expected to find in the Garden that which they do not ordinarily see elsewhere. When one is not expecting to find, in that which belongs to Princes, the greatest efforts of Art, one hopes to see things that are at least uncommon," he wrote, "& yet often one criticizes what one desired—novelty (*Garden at Monceau*, 15)." Nonetheless, Carmontelle enlisted novelty in creating his landscape—though it was a novelty already imbued with meaning in the fashionable, elite circles of Paris. The novelty of the exotic and that of the Picturesque were combined to create a delightful experience for the visitor to the garden.

A perambulation through Monceau took one back in time to antiquity and around the globe (from Europe to the Near East and on to Asia). But there was no apparent logic to the journey; no apparent rationality to the chronological, geographical, and aesthetic arrangements of the scenes and sets. And they *were* sets and scenes through which the visitors walked: Carmontelle maintains no pretext that the garden he designed was anywhere but in eighteenth-century France. While the *View of the Tartar Tent* includes Turkishly garbed figures and the Chinese merry-go-round (brass-ring carousel) is staffed by servants in Chinese-inspired costumes, the visitors they tend are unmistakably, even emphatically, eighteenth-century elites indulging in the illusions created by Carmontelle and the duc de Chartres.

To summon the exotic, past and present, Carmontelle enlisted the obligatory *fabriques* and follies of the

eighteenth-century Picturesque garden. Along with the antique temples and tombs, there could be found an Egpytian obelisk, a Tartar tent, a Turkish minaret, and a Chinese bridge, as well as a quaint peasant farm and mill. The overall effect, celebrated in the painting of Carmontelle handing over the keys to the garden to the duke (fig. 43), seemed illogical to some: Thomas Blaikie (1750–1838), the Scottish gardener brought in eventually to take over the design and planting of the garden, wrote in his diary in 1777 that "the Garden is a confusion of Ruins temples & crowded one upon another, in one place you see a Gothic ruin just by that a Grecian and next a Chinese temple or Pavilion finely quilted, which makes a most singular contrast in so small a compass; there hothouses are not well contrived."[3] In 1783 he commented along the same lines that "those Gardens that had been done by Mr. Carmontelle Architect to the Duke de Chartres at a very great Expence where there was Monuments of all sorts, Countries and ages but placed in such a manner that from every part there was a confused Landskipe, for there were adjoining Chinese & Gothic buildings, Egyptian Pyramids joined to Italian vinyards, the winter Garden adjoining to the Hothouses, was more Beautiful than Elegant; the whole was a Small confusion of many things joined together without any great natural Plan, the walks Serpenting and turning without reason, which is the fault of most of those gardens done without taste or reason."[4]

And yet, of course, the encyclopedic hodgepodge that Blaikie found lacking in taste or reason was exactly what Carmontelle had intended: the surprises that would result from the jarring journey from ancient Rome to eighteenth-century France or the passage from China to Turkey. But what role did plants play in the creation of these scenes? An initial perusal of the engravings in Carmontelle's volume brings a familiar frustration to those looking for concise botanical or horticultural information, whether in eighteenth-century gardening texts or engravings. One might be tempted to conclude that the vegetal was not important to the realization of the Picturesque. Eighteenth-century garden theorists seldom specified the plants to be used in their designs, though a few dictates can be gleaned from the gardening texts of the era. In his *Essay on*

Gardens (*Essai sur les jardins*) of 1774, Claude-Henri Watelet made it clear that plants—flowering and otherwise—had to be freed from the absolute regularity and regimentation that had governed the gardens of Louis XIV and Louis XV.[5] But Watelet was only specific about what plantations should not be. More helpfully, René-Louis de Girardin identified in his *An Essay on Landscape* (*De la composition des paysages*, 1777) five roles that plants should play in the landscape garden.[6] They served to create or frame perspectives. They could hide "disagreeable objects."[7] They could create illusion by hiding the termination of garden spaces, suggesting a perpetual continuation beyond the plantation.[8] And they could give "an agreeable outline" to surfaces lacking contour.

The lack of botanical precision in images of seventeenth- and eighteenth-century gardens can be equally frustrating: what, exactly, did Carmontelle plant in the Jardin de Monceau? What kinds of trees did Hubert Robert depict sprouting out of his ruins? It is often difficult, if not impossible, to discern. Such imprecision flourished even as scientific exactitude was growing in contemporaneous botanical paintings and engravings, which would play key roles in the identification and classification of plants.

Large-format images of trees or shrubs in the wider landscape tended to be less specific. Yet impressive examples exist that demonstrate the ability of painters and engravers to achieve both great expression and botanical accuracy in such images. Accurately rendered botanic specimens could also be depicted in purely imaginary settings. The engraver Hercules Segers in his *Mossy Tree* and *Two Trees, An Alder and a Beech*, used inventive techniques to produce expressive and specific specimens. Similarly, Segers's *The Large Tree* (or *Great Tree*), ca. 1628–29 (fig. 44), is an image that clearly depicts an oak tree, but also one situated in the social history of its locale. One knows instantly that the ancient oak is not only the visual center or landmark of the peasant village where it grows, but also an important part of life in that village, as evidenced by the wooden fence that has been constructed carefully around its base. Work and play go on about it.

Accurately rendered botanic specimens could also be depicted in purely imaginary settings. When Louis-Claude de Chastillon illustrated the Chinese hawthorn (*aubépine de Chine*; *Crataegus pinnatifida*) in his *Mémoire pour servir à l'histoire des plantes* (1676) for the Académie royale des sciences, he could not resist depicting it in a style reminiscent of Chinese landscape painting (fig. 45). The choice amounted to a fascinating demonstration of cultural and aesthetic slippage.

In short, plants could be used to evoke time and distance, as well as the familiar or the exotic. So how were they deployed in the time-traveling tour of the exotic offered by Carmontelle in the Jardin de Monceau? Carmontelle identified very few plants and trees by name: *sycomores* (sycamores), *platanes* (plane trees), *ébéniers* (ebony), *peupliers d'Italie* (Lombardy poplars), *cypress* (cypress), *marronniers* (chestnuts), *tuyas* [*thuyas*] *de la chine* (Chinese arborvitae), as well as *vigne de Judée* (bittersweet nightshade), *lilas* (lilacs), and *sureaux* (elderberries). It is a short list. And yet plants, examined in context of the views created by Carmontelle, mattered. As one traversed the garden

FIG. 45 *Chinese hawthorn*, by Louis-Claude de Chastillon, 1676.

(and traveled around the globe) from one exotic locale to the next, the plantations helped to set the scene. In plate VI, *View of the Ruins of the Temple of Mars*, scraggly plants sprout in the cracks of the crumbling pediment, while a dead tree trunk leans precariously towards the ruins. The unmanicured plantations accentuate the ruination and passage of time communicated by the crumbling Roman temple. Conversely, the *pavillons françois* depicted in plate XIII and the carefully trimmed hedges and trees that frame them draw on contemporary French formal-gardening aesthetics. The style of planting depicted in each engraving was intended to echo or emphasize the *fabriques* it contained. For example, the Wood of Tombs (plate XII), as David L. Hays has observed, is planted with trees associated with death (Lombardy poplars, sycamores, cypress, plane trees, and Chinese arborvitae) that provide a fitting setting for the tombs.[9] And the plantation is depicted as dense and dark, thereby contributing to the somber tone.

In other scenes, Carmontelle used plants to call attention to exotic features or distinctive shapes in the garden. For example, the rounded trees growing on the Isle of Rocks in plate VII echo the curvature of the sky sculpted by the blades of the Dutch Windmill. More prominently, the pillar and obelisk jutting into the sky in the *View of the Circus or Naumachia* (plate XI) are echoed by the Lombardy poplars and other trees growing around the pool. The architecturally "exotic" shapes thus reverberate in the shapes of the trees.

Plants were also used to introduce color into the garden. Carmontelle indicated that the Turkish Tents shown in plate XVII were constructed in red-and-white and blue-and-white striped damask and were planted "in a grove bordered with flowers" (*Garden at Monceau*, 23). The White Marble Temple (plate XV) was enveloped by trees; their green foliage, as Carmontelle pointed out, made a piquant contrast with the stone (*Garden at Monceau*, 22). The plate shows flowering shrubs—again not identifiable, though perhaps roses—in front of the temple and at the base of the trees. The effect of the marble gleaming from among the trees and ornamented in the foreground by flowers would have been quite colorful and striking. The separate blue, pink, and yellow hexagonal flower gardens near the French Pavilions were similarly decorative. Hays has suggested that they were given a Masonic color scheme deliberately, and that the garden can be interpreted as an expression of the duc de Chartres's affiliation with Freemasonry.[10]

Plant color and shape were enlisted together to enhance the sets or scenes. In plate XIV, *View of the Tartar Tent*, the squat tent is echoed by compact, rounded, flowering shrubs. The plumed ornament on the hump of the camel juts into the sky like the Lombardy poplars on either side of it. Behind the animal is a colorful floral border that contrasts with the striped fabric of the tent. The scheme demonstrates the use of plants to achieve a particular exotic aesthetic. But once the management of the garden was handed over to Scottish gardener Thomas Blaikie, the exotic would be achieved increasingly through the incorporation of rare domestic and/or imported plants.

Blaikie began to work for the duc de Chartres as early as 1780. At Saint-Leu-Taverny, the Château du Raincy, and Monceau, he took credit for righting what he frankly saw as messes. He wrote in 1783, "After I changed most of those Gardins and destroyed most of those walks which I thought unnecessary or unaturel, continued likewise the different changes of the Gardins of St. Leu; those changes wrought a great effect upon those Gardins and upon the Mind of the Duke."[11] Blaikie touted his ability to maximize the features and attributes of the landscape while also transforming its flora.

The opinionated Blaikie brought with him his own Picturesque prejudices from the British Isles: with rare exceptions, he did not consider any of the gardens that he encountered well designed, well ordered, or well managed. But his diary reveals him to be an avidly curious gardener, interested in exotic, and especially American, plants. He noted the condition and quality (or lack thereof) of the Parisian gardens and collections he visited and the knowledge and curiosity of their owners. In addition, he offered insights into the state of botanical curiosity and elite plant collecting in and around eighteenth-century Paris and bore witness to the fact that plants were themselves used creatively to excite curiosity and surprise. In 1778, when he went to Versailles with Gabriel Thouin (of whom he said "if not a great Botanist is a good companion"),[12] he stopped at the gardens of Lemonier and reported that "here is a fine collection of American plants and in fine Order, the healthiest I have seen anywhere in France."[13] He also "went and saw the Gardins of the Chevalier de Mustell . . . who has a good selection and many scarce plants," including American bog-species for which he had "a good method of overflowing [irrigation]."[14] He spoke highly of the work of Jean-Marie Morel for the duc d'Aumont at Guiscard (ca. 1775), and admired the garden at Ermenonville, but critiqued Hubert Robert's work at Lavalle, writing that "however fine his ideas is upon canvas yet upon the ground they are without judgements as it is not astonishing he knows nothing of the effects of trees nor their color after a few years growth."[15] He found the king's nursery "at Noisy near Versailles . . . very badly in order," and when he visited the collection of Abbé Nolin, "inspector of all the King's nurseries" at Roulle (at the edge of Paris, near Monceau), he discovered that "this Gentleman is not so much of a Botanist as should be emagined to occupy such a place," but "his American plants is very healthy."[16]

As Blaikie began to work on the transformation of the duc de Chartres's gardens, he sought sources for the exotic (including American) plants that he wanted. He frequented "Henry's Nursery," which "they look upon as the first about Paris," but was disappointed: "he has got a fine Catalogue but hardly any scarce plants, although they are marked he does not know them." He "does almost (all) his business in the Public house," Blaikie added; "although I went several times to him I always found him at his bottle, was obliged to mark out the plants and pack them up myself and at most of the nursery men where I have been about Paris it is much the same. What a difference from England!"[17] He acquired some specimens from Thouin and Nolin, including Lombardy poplars. But other sources would be necessary in order to enrich the garden.

In January 1781 Blaikie sent Archibald, an English gardener who worked with him, back to England "to buy green house plants for the Duc de Chartres at Monceau." Archibald "returned the month of March with a fine collection." The plants came from a range of exceptional sources, Blaikie reported: "Received a large quantity from Sir Joseph Banks of those brought home by the Discovery Ships which went with Captain Cook the 10th Feby 1776 and likewise some seeds from Mr. Morison with a great many other evergreen tree seeds." Together, he wrote happily, they "make a compleat Nursery; the only thing wanted was a Hothouse for several sorts of seeds from the friendly Islands and Ottahiette [Tahiti]."[18] The duke himself was in England in 1785 and visited different Gardens about London. Blaikie, on the English side of the channel as well, "went and saw Mr. Aiton, there collection very fine Many rare and New plants; bought several new sorts for the Duke de Chartres."[19]

To acquire the American plants that he wanted, Blaikie emulated the actions of his British counterparts and turned to American nurserymen. Archival sources show that he and the duc de Chartres purchased American specimens from John Bartram and his Philadelphia cousin, the seedsman Humphry Marshall. In 1785 Marshall published his *Arbustrum Americanum: The American Grove, or, An Alphabetical Catalogue of Forest Trees and Shrubs, Natives of the American United States, arranged according to the Linnaean System*. The volume was one of the first books published on American plants by an American. There are clues in Marshall's text that he was targeting an international clientele: "The *foreigner*," he wrote, "curious in American collections, will be hereby better enabled to make a selection suitable to his own particular fancy. If he wishes to cultivate timber for œconomical purposes, he is here informed of our valuable Forest Trees: if for adorning his plantation or garden of our different ornamental flower shrubs."[20]

Marshall conducted business with a range of French customers—first and foremost French officials posted in the United States. Marshall supplied seeds to the French government via requests placed by Conrad Gérard and the marquis de Marbois, diplomats posted to Philadelphia, and Hector St. Jean de Crevecoeur, a former French military officer who had established himself in America. But Marshall also catered to private clients. Stephen Girard, a French sea captain who had settled in Philadelphia, served as a go-between, writing to Marshall that he had been "requested by a Gentleman in Paris to apply to you for the following plants,"[21] which included magnolias, kalmias, franklinias, and azaleas. Étienne-Thomas de Maussion, *intendant de Rouen*, ordered plants and seeds—presumably for the Château de Jambville, which he had purchased about 1775. In 1787 Marshall received a large order from Adrienne-Catherine, comtesse de Noailles de Tessé, Parisian *salonnière*, aunt to the Marquis de Lafayette, and passionate gardener,[22] who was constructing a Picturesque garden at the Château de Chaville.

Marshall also received an extensive order via Philadelphia merchant John Vaughan from Louis-Philippe-Joseph. The order is undated, but Louis-Philippe is identified as the duc d'Orléans, therefore it could not have been placed before 1785 when he succeeded to the Orléans title on his father's death.[23] The list does not indicate for which of the ducal gardens Marshall was asked to furnish plants, but the order shows that the duke had come around to Blaikie's way of thinking: the garden of *addition* now featured the addition of plants.

In the order the duc d'Orléans requested both live specimens and seeds, and he included detailed instructions on how they should be packed and shipped. Among the many different species requested were varieties of sweet gum, sassafras, magnolia, American plum, viburnum, alder, birch, maple, and oak, as well as gardenias, mock oranges, mountain ash, American olives, summersweet, franklinias, azaleas, blueberries, rhododendrons, mountain laurel, plus birthwort and trumpet vine. Seemingly intended for different destinations, each box of seeds contained a broad variety of itemized species. Collectively the contents would have contributed a wide range of colors and shapes to the landscapes for which they were destined.

The writer of the list added that "we ask of Mr. Bartram that he send 4 to 5 boxes, of the sort he has already sent to France." The request implies that this was not the duke's first order or merely that he had seen plants sent to others. That the present order was filled is suggested by another document in the Archives nationales, a "Listes des Graines et plants arrives d'Amerique en 1788 et semes dans les differents jardins de Monseigneur le duc d'orleans." The seeds on the list included over two hundred different varieties, many of which appeared on the order that was transmitted by Vaughan. Among the live plants listed were hundreds of additional trees including pines and cedars, and there were also "two cases of plants and two of seeds from Bartram and Marshall." One of the four cases—whether of plants or seeds is unspecified—was furnished by Benjamin Franklin. The cost of the items on the list totaled 3,146 livres—a considerable sum.[24]

Of course, Carmontelle's publication predates Blaikie's work at Monceau and the transformation under the latter's leadership. But the garden of the duc d'Orleans would continue to be a destination that delighted its visitors. Their curiosity and amazement would increasingly be piqued by the plants themselves. Hints of Blaikie's work are visible in the engravings of Monceau by Georges-Louis Le Rouge, who dedicated two images of his *Jardins anglo-chinois, ou Détails de nouveaux jardins à la mode*, to representations of the Berceau d'Hiver at Monceau, a complex structure that included a greenhouse and grotto where rare plants could be enjoyed in the off-season (figs. 46 and 47). That the space was intended for visitors' use is reinforced by the depiction of a platform on which musicians would perform (seen at the top of fig. 46). The addition of the complex suggests that plants from America and elsewhere around the world were coming to play a growing role at Monceau: they allowed visitors to indulge in the pleasure of "not seeing the same things often."

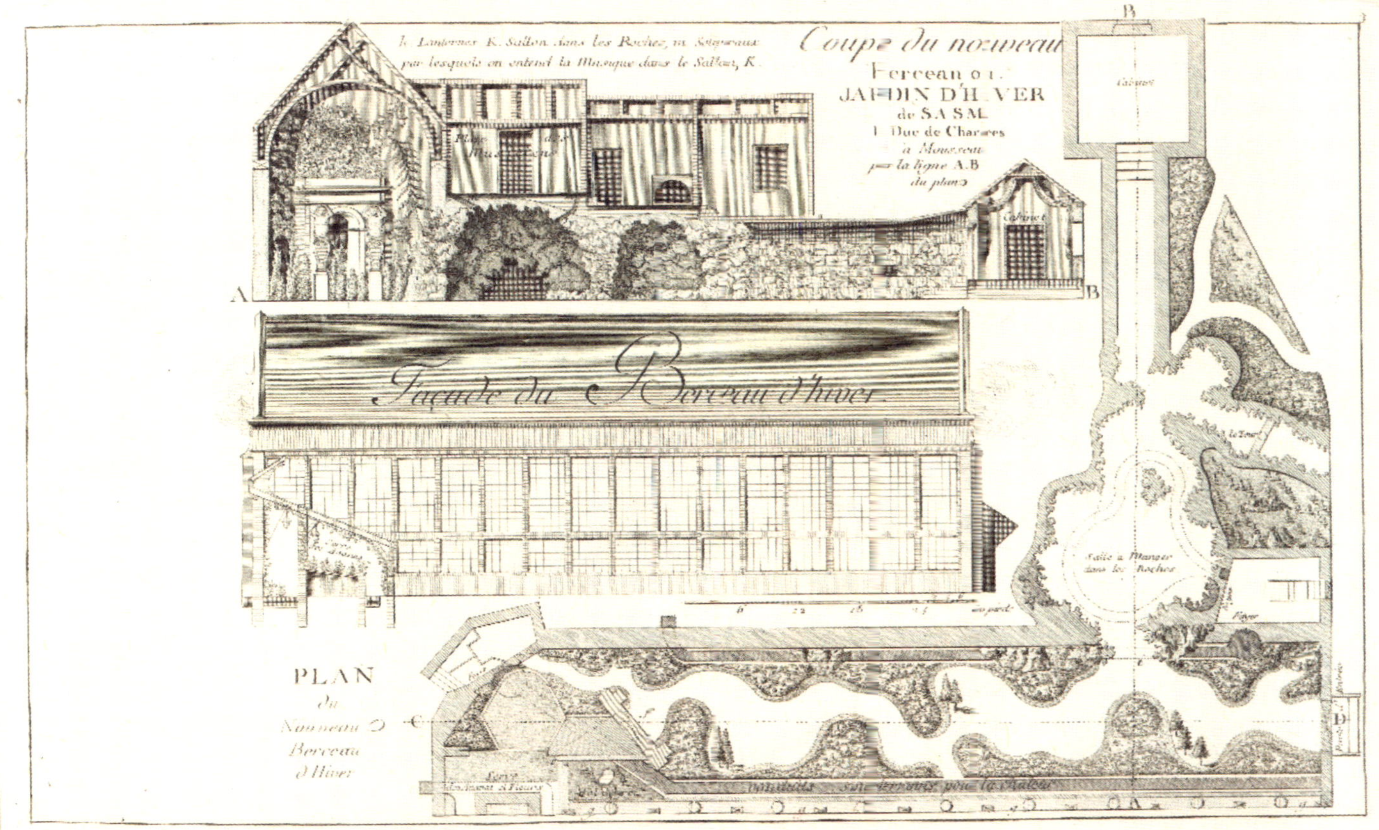

FIG. 46 Blaikie's Berceau d'Hiver at Monceau, engraved by Georges-Louis Le Rouge, ca. 1783.

FIG. 47 Longitudinal section of the Berceau d'Hiver, 1783, from another engraving by Le Rouge.

FIG. 48 A celebration at the Colisée in 1772, by Gabriel-Jacques de Saint-Aubin.

MONCEAU, THE *MÉMOIRES SECRETS*, AND THE COLISÉE: THE REINVENTION OF THE DUC DE CHARTRES *À L'ANGLOIS*

Gabriel Wick

Uring the summer of 1773, a series of newsletter articles that circulated in manuscript described the transformation of the property at Monceau owned by the young duc de Chartres from an introverted *petite maison* into an expansive, extroverted landscape garden incorporating the latest trends in design, materials, and engineering.[1] The first such article, dated June 21, 1773, read:

> My lord the duc of Chartres has taken on an extraordinary taste for Sieur Comus . . . who has developed slights of hand to a superior degree and has reduced this art to a few subtle principles. The prince is taking lessons, and . . . has had a small house built for himself above the faubourg du Roule, where he practices these games and other amusements of his age.[2]

A week later a second article appeared that gave greater detail:

> The garden of M. Boutin, Tivoli . . . has inspired the duc de Chartres to create something similar, which is to say yet more magnificent and worthy of his Serene Highness. It is an enchanted resort, which makes the prince's orgies all the more delectable. It is Carmontelle, an enlightened amateur of the arts who cultivates them himself and is already well known for a collection of proverbs, a talent for caricature, and some very realistic society portraits, who directs these works. Amongst other curiosities is a steam pump that raises water in order to produce a river from wells.[3]

In August a further report described how the new structures on this estate would make use of the latest in building materials:

> Loriot, a machinist known for his numerous inventions and machines, has discovered a mastic that equally resists both rain and heat . . . His Majesty has purchased this secret with a contract of one hundred thousand francs . . . M. le duc de Chartres, having need of Loriot for his *petite maison*, made a request to the contrôleur-général that he be allowed to work on the project, which was duly permitted. Loriot's mastic can take the place of tiles or slate for the covering of a house.[4]

The Comus to which the first article refers was Nicolas-Philippe Ledru, a scientist who was also a famous impresario. His physics cabinet was one of the most popular destinations in Paris in the 1770s. The steam pump at Monceau, to which the second article alludes, was remarkable for being one of the very first in Paris. The machinist mentioned in the third article, Antoine-Joseph Loriot, had rediscovered an ancient Roman technique for mortar or cement that allowed structures to resist the elements effectively.[5] The editor of the publication in which these articles appeared, the *Mémoires secrets*, was Mathieu Pidansat de Mairobert, a journalist and propagandist, who served as secretary of armaments within Chartres's household. Through such tantalizing reports the readers could project themselves mentally into the private retreat of one of the kingdom's wealthiest and most exalted subjects. Thus, although Monceau was ostensibly conceived as a private space, it seems to have been designed to capture the attention and imagination of the public at large.

The expansion and transformation of Monceau by the duc de Chartres in the first half of the 1770s is notable as one of the earliest and most prominent uses in France of the English-inflected, irregular style of landscape gardening "*à l'anglois*." Monceau, however, broke new ground not just in terms of aesthetics but also in the ways in which it allowed Chartres to show himself to the people of Paris. In this the garden was something of a hybrid between the traditional urban or peri-urban princely or aristocratic gardens such as the Palais-Royal or the promenades of the Temple, Luxembourg, and the park of Saint-Cloud, which had long served as popular spaces of recreation, and a new category of urban public space, also recently introduced from England, the commercial pleasure garden (called in France the "Wauxhall" or "Vauxhall"). Specifically, the grandest and most ambitious of the commercial pleasure gardens in Paris, the short-lived Colisée (which operated from 1771–77 on the Champs-Élysées) offered Chartres an apposite model for capitalizing on his celebrity and creating for himself a unique form of visibility.

Unlike the English landscape gardens it emulated superficially, which were located, for the most part, far from urban centers, Monceau sat within easy walking distance of the city. It was only 1.6 miles from the Palais-Royal or half a mile from the Colisée. When its owner was not in residence, Monceau, like many aristocratic gardens, was open to the visits of elites or those judged to be *honnêtes gens* (respectable persons).[6] All was not lost for those who could not gain admission, however, because the ha-ha that ran along parts of the garden's northern, eastern, and southern edges, which allowed the duke and his intimates to admire the surrounding countryside, also allowed the public to gaze in. As Emmanuel de Croÿ-Solre, duc de Croÿ (1718–84), recorded in his personal journal in June 1775: "I was able to see with the greatest ease, from the outside, walking along the fields, the singular garden of the duc de Chartres; it's the true Chinese tone, and an ornament of China made real."[7] Given Monceau's comparatively small scale, many of the garden's most important features were visible from the exterior. Carmontelle's plan of the garden from 1779 shows that visitors at the edge of the ha-ha were only 40 feet from the Naumachie, 75 feet from the Temple of Mars, and 150 feet from both the *jeu de bague* and the Minaret.

Croÿ was hardly alone in viewing Monceau from without. Thiéry's *Guide des amateurs* informed readers that the printer La Fosse had engraved seventeen perspectives of the various views the visitor could have during "a promenade around the open perimeter of the garden."[8] The commercialization of such small-format prints (presumably with the duke's sanction) alongside Carmontelle's lavish and costly monograph demonstrates that the garden captured the interest not only of Chartres's élite peers but also of a public of far more modest means. Indeed, it was this degree of exposure that damned Monceau in the eyes of the prince de Ligne: "One is isolated in the midst of a vast plane that is boring to look at. One is seen from all sides. This was supposed to be a garden: but it is everything but that."[9]

Chartres had much to gain from this innovative form of visibility. From April 1771 until January 1773 he was formally forbidden from attending court. In this he was hardly alone: exile and disgrace had become something of a leitmotif in the last years of the reign of Louis XV.[10] In December 1770 the king had exiled his powerful first minister, the duc de Choiseul, to the latter's country estate of Chanteloup. In a remarkable break with tradition, however, many of the most powerful figures at court showed their continued support for Choiseul by visiting him there. In April 1771 the chancellor Maupeou exiled the magistrates of the *parlements*, the kingdom's highest judicial authority, to distant and inhospitable corners of the kingdom. When all but one of the princes of the blood signed and published a letter making their objections to the crown's actions known, Louis XV sent them away from court.[11] While the other princes of the blood engaged in such acts of resistance reluctantly, Chartres and his renegade uncle, Louis-François, prince de Conti (1717–76), assumed prominent roles as figureheads of the controversial opposition movement, the so-called *parti patriote*.[12] The time Chartres spent away from court was quite short: in January 1773 his father, the duc d'Orléans, insisted that he join him in making peace and returning to court. Chartres, however, made it clear to the king and everyone else that he was still determined to resist the new government. He informed the king that he would visit Choiseul in exile, a gesture tantamount to open rebellion. After January 1773 his rebelliousness would assume a more gestural and performative character.

For Chartres, appearing in public was an essential means of maintaining his prominence as a figurehead of opposition. These appearances, closely followed in the press, served to reassure his allies that he was not at court and remained faithful to their cause. Luckily for Chartres, the house of Orléans, of all the princely dynasties, had always maintained an unusually close relationship with the people of Paris. As the *Mémoires secrets* noted, Parisians tended to regard this branch of the royal house that resided in their midst—in the Palais-Royal—with a special affection.[13] The Palais-Royal was home to the Opéra de Paris, and its gardens as well as those of Saint Cloud were popular resorts, and meeting places for journalists and observers of the political scene. Chartres's own personal celebrity ensured that the public would always hunger for, and seek out, details of his private life.

Although the princes had much to gain from political activism, the king wielded an enormous degree of financial power over them. Governorships, pensions, and property were his to give or withhold. Everything from travel to real-estate speculations to marital projects depended upon the king's assent. In extreme cases he could consign the princes to distant country estates where they would effectively be removed from political life altogether. Thus even princes of the blood needed to temper their actions with a degree of restraint and nuance. An account from one manuscript newsletter from the summer of 1773 described how the prince de Conti, Chartres's role model in all things political, was advised that his daily habit of walking along the boulevards near his palace of the Temple without an entourage, dressed simply, was exciting the suspicions and anger of the government. He was warned against speaking with the people he encountered there and making such shows of affability.[14]

The arrival of the Vauxhalls, however, presented a new and unprecedented opportunity for apparently spontaneous and unstructured interactions between grandees and the middling public. As John Goodman notes, the possibility of rubbing shoulders with the high aristocracy was an important part of the attraction of such venues.[15] For the princes such places offered important opportunities to present themselves to the public: they could do so passively under the innocent guise of taking in a simple entertainment. The Colisée—with a capacity of forty thousand people by far the largest and most ambitious of these establishments—had opened in May 1771, only a month after the princes' departure from court.[16] Choiseul was rumored to have had a significant investment in the project.[17] His household architect, Louis-Denis Le Camus, had designed it.[18] Much of the Colisée was taken up by a vast reception court surrounded by terraces that provided the public with an excellent vantage point on the latest arrivals (fig. 48). Given that here, as at the theater or opera, notables were applauded (or not) upon their arrival, such venues offered a very public measure of one's standing and popularity. As the *Mémoires secrets* noted, Chartres was warmly received at the Colisée after his departure from court, while his cousin, the comte de La March, the only prince to remain at court, was received frostily by the assembly.[19]

Despite the patronage of Chartres and other leading dynasts, the Colisée's commercial success was far from certain from the beginning. As one might expect from a project planned by a clique of aristocratic investors, certain logistical and practical aspects had been largely neglected. The cost of illuminating its lofty interiors alone was rumored to consume its evening receipts. The garden's principal feature, a vast pool surrounded by seating and a colonnade, was intended for water jousts, but its presence rendered the space unsuitable for any other type of performance (fig. 49). Worse, the pool's waters quickly went stagnant, and a rank odor pervaded the space. In a very short time, coverage of the Colisée in the

FIG. 49 Nocturnal water spectacle in the Colisée in 1772, by Saint-Aubin.

FIG. 50 The Naumachia of Monceau in 1778, by Saint-Aubin.

manuscript newsletters went from recording the appearances of the princes of the blood and younger members of the royal family to lampooning the increasingly desperate attempts made by its investors to attract the public.

With Monceau Chartres was able to transpose the unique form of visibility afforded by the Colisée to a venue that he controlled completely. As is evident in Carmontelle's engravings, Monceau was less a garden and more a permanent spectacle. It could even boast of its own cast, as the duke's servants were obliged to don exotic costumes and parade with his collection of rare animals—most notably his dromedary. Like the Colisée, Monceau's most monumental feature was its pool surrounded by a colonnade: the Naumachia. Such pools were associated with the villas of the ancient Roman emperors, where they served as the setting for staged maritime battles. The Naumachia at Monceau was situated in the most publicly visible corner of the garden, where the ha-has that ran along its northern and eastern edge met (fig. 50).

The Naumachia's evocation of maritime glory may have held a personal resonance for Chartres. Starting in 1773 he had engaged in a determined campaign to gain the prestigious and profitable charge of Grand Admiral of France. Even though his hopes for this honor quickly dimmed, Chartres remained nonetheless convinced that his most likely path to glory and distinction lay with the navy. In 1775, in direct contravention of Louis XVI's orders, he joined the navy as a seaman without rank. Such romantic and unconventional behavior only added to Chartres's popular reputation and was a source of immense personal pride.

Other elements within the garden seem intended to illustrate the duke's modern and iconoclastic nature. As David L. Hays notes, the armillary sphere that occupied a highly symbolic position in front of the principal entrance evoked Chartres's rank as titular head of France's Freemasons.[20] The garden's Wood of Tombs evoked his sensibility and emotionality. One somewhat enigmatic feature seems far more controversial: near the armillary sphere stands what Carmontelle describes innocuously as a perch for target practice. In fact, however, this object, whose proximity to the house would seem to make it impractical for shooting, bears a striking resemblance to a liberty tree topped with a pileus or Phrygian cap. As J. David Harden observes, the helmet-shaped, felt pileus was worn by manumitted slaves in ancient Rome and had been widely adopted by Dutch, English, and American patriotic movements in the 1760s and 1770s as a symbol of their demands for liberty and their resistance to tyranny and despotism.[21] Its presence here, along with the armillary sphere, seems to have been meant to alert the visitor to the possibility that, interwoven with the apparent frivolity and playful illusion of the gardens, there were potent politically and ideologically charged meanings.

By the autumn of 1773, however, Monceau's coverage in the manuscript newsletters was largely eclipsed by the news of the birth of a son (the duc de Valois, the future king Louis-Philippe) (fig. 51). The *Mémoires secrets* noted approvingly that, in an unprecedented gesture, the duke and duchess welcomed the public to come into the palace and gaze upon him.[22] Another article indicated that the child, in a basket and free of swaddling clothes and other shackles of old traditions, was visible daily in the private garden of the Palais-Royal, which was separated from the public promenade only by an iron grille.[23] Thus even the couple's fulfillment of their most traditional obligation, the provision of a male heir, was framed as a media event, and one that showcased their modernity and relatively egalitarian principles.

Other developments were to further erode Monceau's key place in the construction of the duke's public identity. As we have seen, in the second half of the 1770s, when irregular-style gardens were becoming increasingly commonplace, Chartres had exchanged such picturesque enclosures for adventures on the high seas (which were to end ulti-

FIG. 51 Louis-Philippe, duc de Valois, in his cradle in 1774, by Nicolas-Bernard Lépicié.

mately with humiliation at the battle of Ushant).[24] At a time when French courtiers were normally loath to leave the Île-de-France, Chartres and his wife developed a remarkable interest in travel in order to experience, as it were, the Picturesque landscape writ large. In 1776 the duchess accompanied Chartres to meet his ship at Toulon and then made her way through Italy. In 1778 she stole away from Marly in the company of the princesse de Lamballe to explore Brussels and Amsterdam. For some days the pair evaded the men dispatched to bring them back.[25] That year Chartres commissioned a portrait of his wife by the painter Joseph-Siffred Duplessis in which she is shown barefoot posed romantically amidst the rocks on a desolate seacoast, while the vessel the *Saint-Esprit* bears her husband off to the horizon and to battle (fig. 52). Nothing could be further from the contrived and contained Picturesque evocations of the garden of Monceau than this depiction of untamed wilderness and personal risk. The next year, 1779, Chartres would take his search for adventure even further in attempting to board a ship against the orders of the king to go and fight with the American insurgents.[26]

By the end of the 1770s it was clear that Chartres had developed a taste for adventure that went well beyond gardening. For all its apparent playfulness and fantasy, we can see in Monceau, as well as in Chartres's involvement in the *partie patriote* and the *Fronde des princes*, the first stirrings of political ambition and of innovative and sophisticated forms of self-representation that would flower in the 1780s. In this sense we can understand Monceau not just as a garden or a playground but as a space of experimentation and transformation where Chartres could recast his inherited status and position into something far more powerful and, for the absolutist monarchy, quite fatal.

FIG. 52 The duchesse de Chartres and the departing *Saint-Esprit*, carrying the duc de Chartres to the battle of Ushant, by Duplessis, 1777–78.

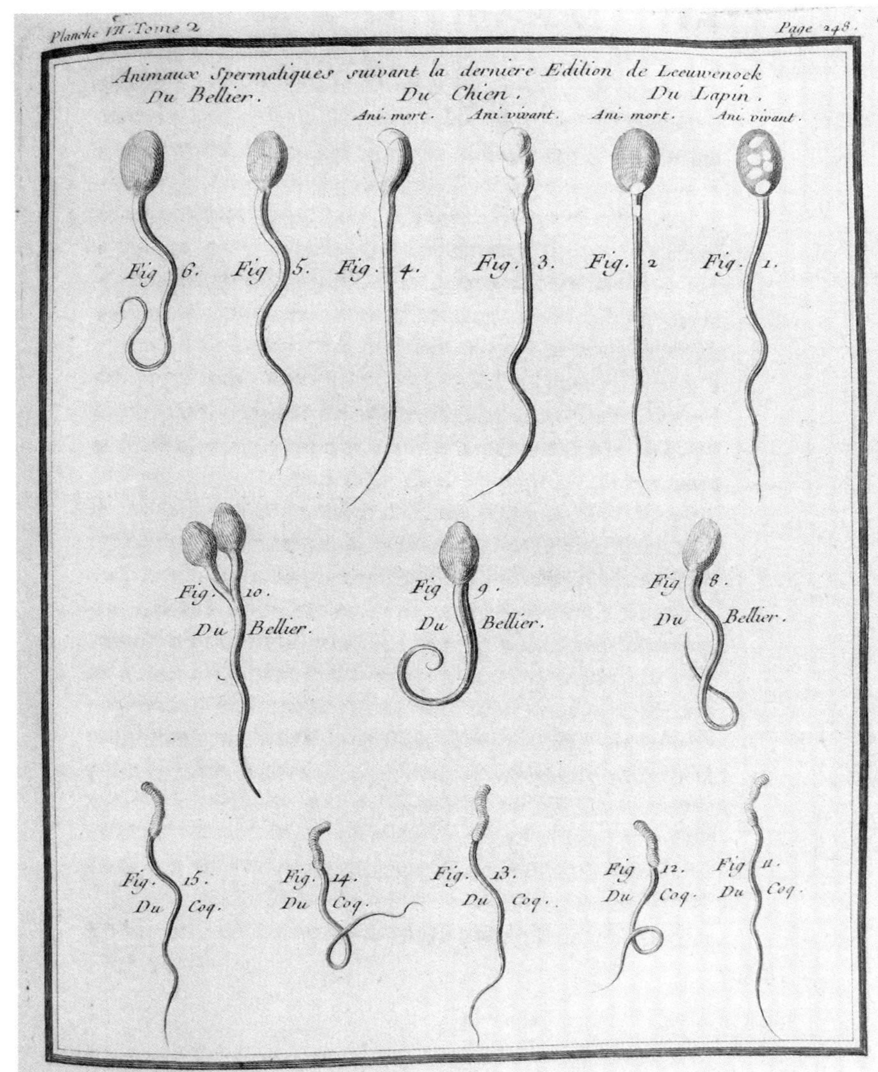

FIG. 53 Sperm cells, from Buffon's *Histoire naturelle*, 1749.

A SPADE IN THE GARDEN OR A GARDEN IN SPADES: LOUIS CARROGIS DE CARMONTELLE'S *JARDIN DE MONCEAU* OR THE GARDEN AS DRAMATIC PROVERB

Joseph Disponzio

CARMONTELLE'S garden at Monceau is not your typical French Picturesque garden. As his lavish folio illustrates, Monceau is a "land of illusion"—a tableau vivant of garden elements made material in earth, water, plant, and rock. It is a fantasy landscape where one can ride a magic carpet of wild flowers from China to Turkey, Egypt to Holland; be hunter, farmer, milkmaid, or chemist; ride on a camel, row a boat, descend into the underworld and return. A world where winter is made summer and summer is made splendid; a potpourri of garden elements whose themes are culled from mythology, theater, science, the world at large, and Carmontelle's restless imagination. There are no tiresome walks in Monceau; no unpleasant adventures. There is only one requirement in Carmontelle's ancien régime, plein air fun house: be gay!

There is much to be considered in Carmontelle's *pays d'illusion*. It is the unique creation of an ingenious man more gifted in contrivance than composition, more comfortable with satire than sincerity, more interested in illusion than reality. Frivolous at enormous expense, Monceau is a private joke filled with irony. As a garden it was an exemplar of the French *jardin anglo-chinois*. Yet it is more parody than conceit, a critique, if not a negation, of gardening art and its theoretical debate. Monceau is also a satirical comment on fashion and society—the society that adored and sustained Carmontelle but could not accept him.

While Monceau is undoubtedly his greatest achievement, Carmontelle was better known during his lifetime for pencil portraits of society's elite and writings for theater—especially a genre known as the Dramatic Proverb (*Proverbe dramatique*). His portraits are innocuous enough; they are straightforward, minutely detailed representations that neither flatter nor degrade the sitter. His writings, however, are not as neutral. His plays and Dramatic Proverbs are filled with polite humor and pointed sarcasm written in the eighteenth-century spirit of persiflage, which is characterized by light ridicule and laser-pointed raillery; ironic and inane stream-of-conscious connections; mockery, disingenuousness, and gibberish. Carmontelle was adept with the form, which gave him an advantage in a society in which quick-witted repartee was required to hold one's ground. Such legerdemain is a double-edged sword, but Carmontelle was expert with his creative épée.

Implicated here is another side to Carmontelle: one less documented, lesser known, and less well understood. In his time he was the undisputed "painter" (read: observer) of society, enjoying a charmed world of leisure. Yet he occupied a subaltern presence within that society, acting a court jester as it were. His primary responsibility was to protect society from its worst enemy, boredom, as Laurence Chatel de Brancion has so aptly noted previously in this volume.

It is his arrière-pensée—the critic's yin to the jester's yang—that informs our interpretation of Monceau. How did Carmontelle reconcile his higher, noble artistic aspirations, which he is known to have had, with his indenture at the court of the house of Orléans, where his job was to amuse and flatter a frivolous society? Given the events that soon would end in catastrophe for most of his audience, it is not without interest that he survived the Revolution and the Terror. That he was part of a libertine, if not liberal, household jealous of princely prerogatives and frequently in opposition to the king may have worked in his favor, but it did not help his patron, the duc de Chartres, who, despite his vote for regicide, himself fell to the guillotine. Judging from appearances, Carmontelle was not an openly political man, but he is known to have engaged in sensitive activities, and he was a vocal critic of the Royal Academy of Painting and Sculpture, the official organ of the arts in France. Unfortunately, too little is known of this side of Carmontelle's life, the counterpoint as it were to his life of outward show without ideology, complexity, or controversy.[1]

For over twenty years Carmontelle served the house of Orléans, the rival family to the Bourbon throne. His titular role was tutor to the duc de Chartres, the future Philippe-Égalité, but in reality he was an indispensable member of the Orléans household, engaged to keep it entertained. During his tenure within this prestigious ménage, his duties ran the gamut from poet to painter to architect to interior designer. Above all, he was an ingenious *artificier*—event planner—serving a world in need of endless amusements.[2]

His attachment to the Orléans court allowed him entree to all segments of *le beau monde de Paris*. Carmontelle knew everyone worth knowing in the ancien régime, from philosophe to artist to critic. Mme de Genlis, the respected chronicler, keen observer, and acerbic diarist of society recalled her friend as being of gentle character, talented, reserved, and loved.[3] Others, however, found him severe, sardonic, and imperious.[4]

How he mastered the rigorous and suffocating etiquette of the time is anyone's guess, yet he became adept at negotiating the era's social norms. Emphasis was placed on grace and manners, which he seems to have had in abundance. As important, in a world of imperative gossip, Carmontelle knew how to keep his polite and amusing, and within the bounds of what was appropriate, acceptable, and permitted. In brief, he abided by the decorum of a society whose canon he had mastered.

Carmontelle entered the Orléans court shortly after the Seven Years War (1756–63). His position thrust him squarely into the most "brilliant and spiritual" court of the ancien régime. At the midpoint of his life, he found his métier—society and the court became the locus and source of his life's work. Over the next two decades he would capture in word and in drawing the essence and detail of a world of wealth, etiquette, pleasure, and boredom.

At the head of the court was Louis-Philippe d'Orléans, great-grandson of Louis XIV's brother, Monsieur, Philippe de France. Fair—he took after his German mother—and fat, he was called "le Gros"; he had little ambition, taste, or intelligence. He surrounded himself with men of letters and artists. Incapable of judgment on his own, he saw things through the eyes of others. He loved the theater and was said to have excelled in peasant roles. He was prone to debauchery—food, parties, and women were his passions. At thirty he had a serious attack of gout for which his doctor prescribed moderation. He eventually died of apoplexy.

His wife, Louise-Henriette de Bourbon-Conti, was his equal in licentiousness. The pair had a scandal-ridden marriage. She was said to walk the gardens of the Palais-Royal at night in search of peasants—so much so that her escapades were immortalized in quasi-pornographic writings of the time. The looseness of their conjugal relations prompted anti-Orléanists to discredit the legitimate birth of Louis-Philippe-Joseph, but Louis le Gros never doubted that the boy was his son.[5]

Louis le Gros's favored mistress was Étiennette-Marie-Perrine Le Marquis, oyster seller turned dancer at the Comédie-Italienne. Le Marquise, as she was known, indulged his considerable weaknesses. He built private theaters for her where she tried to emulate the social theatricals held by Mme de Maintenon in the previous century and Mme de Pompadour in the present. Her favorite dramaturge was Charles Collé, the master of the lewd and erotic *parade*. The court of the duc d'Orléans fell to the limits of acceptability as society began to shun Le Marquise's theater. Pressured to discharge his mistress, he replaced her with Charlotte-Jeanne Béraud de La Haie de Riou Montesson, a woman of remarkable calculation for survival. At sixteen she married the very rich elderly marquis de Montesson. By thirty-two she was a woman of fashion and society, and a widow. She also loved theater and was said to be a great actress in her own right.

Mme de Montesson had a sobering effect on Louis le Gros and brought a semblance of decorum into the Orléans household. Under her guidance the Théâtre des Petits Cabinets was reinvigorated and became the most important private theater of society. Collé and his vulgar plays were banished in favor of a more refined writer such as Carmontelle. He became one of Mme de Montessson's favorite writers and remained so throughout his association with the house of Orléans.

Meanwhile, Carmontelle's duties as tutor to the duc de Chartres must have been a challenge. Born of privilege and wealth, Louis-Philippe-Joseph was raised as a successor to the crown, but his education and rearing stressed courtly behavior rather than responsibility and action. By all accounts the young duke was incorrigible. While intelligent and spirited, he resisted learning and was averse to reading.[6] Entering his teens, he was brought to the celebrated Mme Duthé, known for deflowering young nobles. Mme de Genlis, who called Duthé an "execrable creature," feared for the moral education of the young man.[7] Her fears were justified as Chartres went on to have dozens of mistresses before (and during) his marriage to Louise-Marie-Adelaïde de Bourbon.

They were married in April 1769 with nuptial celebrations continuing throughout the spring and summer. Carmontelle was an indispensible part of the ceremonies, serving as organizer, designer, and director of the wedding spectacles.[8] Indeed, the duties he rendered for the princely marriage were more appreciated than those belonging to his salaried position. He was in demand everywhere.[9]

It was in such a household that Carmontelle thrived. As noted, with his ringside seat on the activities of the ancien régime, he not only observed his era but recorded it both in words and images. Carmontelle's portraits, executed in near microscopic detail, are an invaluable iconographic record of the time.[10] Often his sitters are outdoors, suggesting the importance of the landscape as a romantic, imaginative vehicle for their enhancement. Such attention to natural detail would also become apparent in his *Jardin de Monceau*.

Carmontelle's portraits comprise a *défilé* of men, women, and children, famous or otherwise, of the ancien régime that neither pushes artistic boundaries nor stirs societal norms. Yet their creator was not averse to depicting politically charged subjects. One of his drawings, which became an internationally distributed subscription engraving, was of the imprisoned Calas family.[11] Jean Calas, a prosperous Huguenot of Catholic Toulouse, was accused of murdering his son, who had converted to Catholicism. The *parlement* of Toulouse condemned Calas after a trial inflamed by religious fanaticism and mistrust of Huguenots. Voltaire and Diderot, among others, came to the defense of Calas. The king was petitioned. Eventually Calas was exonerated but not before he was broken on the rack and his family imprisoned and ruined. Carmontelle sketched the family in prison on March 9, 1765.

No less political was Carmontelle's portrait of the Jansenist parliamentarian Henri-Philippe Chauvelin, which was also made into an engraving.[12] Chauvelin orchestrated and led a campaign to expel the Jesuits from France, for which he became a temporary popular hero. Carmontelle also drew a portrait of the Jansenist *contrôleur general*, Claude-Guillaume Lambert.[13]

More substantive, layered, and satirically charged was Carmontelle's literary production, including plays, comedies, dramatic dialogues, and other works for the stage, as well as *mémoires*, critical essays, three novels, and over 130 Dramatic Proverbs. Despite his considerable output and popular success, he was considered a limited talent. Grimm was severe. He acknowledged Carmontelle's "prodigious fecundity" and appreciated his knack for representing characters and writing dialogue, but he found Carmontelle's theatrical work cold, flat, and devoid of artistry. He had talent and ingenuity, but the essential ingredient needed to transform imitation into art was lacking.[14]

Despite the severe criticism, Carmontelle could take consolation in his popular success. He became the undisputed master of the Dramatic Proverb; indeed, the revue *L'Année littéraire* proclaimed him "the Molière of Proverbs." According to Carmontelle, the Dramatic Proverb is a light comedy that illustrates a word or maxim hidden within the text, which the spectators must guess. The intention is to have the audience experience an "ah-ha" moment when the word or maxim is deciphered. As theater it is a slight work of short duration intended for pure amusement, staged in the privacy of the *petit cabinet*, squeezed in before dinner and brought to life by the guests for an evening's entertainment. Its limited convention worked in its favor, and its modest ambition led to its success.[15]

Though his output was decidedly lightweight, Carmontelle himself was not, in his own estimation. His focus and attention to detail seemed incommensurate with the narrative action. For example, he designed and painted each scene to scale and designed each character's costume so meticulously that the level of detail extended to a garment's particular thread.[16] He even developed a color symbolism signifying social class, moral character, and personal status or disposition. Coquettes such as the lemonade seller (*la limonadière*) Marianne, daughter of Mme Mignonette—her name itself a pun—wear lively colors such as red and green. The lovelorn, melancholic, widowed, cuckolded, etc. wear gray or other somber colors.

Carmontelle was also a student of theater history and he seeded his plays and other works with references to past and current French immortals. The authority of Molière is invoked in his proverbs, while his play *Théâtre du Prince Clenerʒow* (1771) paraphrases Diderot's *De la Poésie dramatique* (1768). As a critic he followed the lead of Diderot, critiquing the rigidity of French theater in general, while praising, as had Diderot, the realism, action, and pantomime of Italian theater.[17] The *commedia dell'arte*, to which he owes his concept of theater and sense of timing, was acknowledged in his preface to his first collection of dramatic proverbs.

There is a rub, however, to Carmontelle's success. He was constrained to write for an audience that bored easily, quickly, and habitually. His compromises were great. He could not depend on a cohesive and well-trained acting troupe whose members were accustomed to working together; instead he had to rely on undisciplined amateurs. His works had to be short and simple, making use of easily recognized conventions and stereotypes, such as well-known plays on words; physical, mental, and moral infirmities; amphigory, and the like, so that neither actors nor audience would be taxed with complexity. No one wanted to have to think, be bored, or lose patience, especially before dinner.[18]

For a man of artistic aspirations, this must have been demeaning. Baron de Frénilly recalled how Carmontelle would become enraged over a dropped or misplaced word. He required his prose to be learned as "religiously as the lines of Racine."[19] Racine! Carmontelle was not in any position to scold or make demands. He was very much a domestic in an illustrious household. He knew it and so did his contemporaries. While he held an "honorable" position of sorts, Mme de Genlis recalled that, even at a country seat where strict social etiquette was relaxed, he "didn't have the right to dine with the princes."[20]

From the inseparable distance of class he viewed the excesses of nobility. His job was to amuse, flatter, and exalt the rare qualities and virtues of the people he served. The truth was different, and he knew it. So his theater is filled with infidelities, gambling debts, penury, conceit, vanity, stupidity, and boredom. His favorite targets—the aristocracy, the bourgeoisie, the clergy—are exposed in all their frivolity, cynicism, debauchery, and abjection. He wrote some sincere works without much success. He did ridicule the lower classes, but in general they are treated sympathetically: poor man turned rich seigneur finds being poor better; masters are stingier than maids; the lower class is morally and intellectually superior to the aristocracy.

There is much irony in Carmontelle's written work. He portrays a paradisiacal world in a hellish universe where everyone is in constant search of the next pleasure. It is a contradiction that no doubt escaped the aristocrats whom he served yet he fueled his creativity with their blindness. While Carmontelle was more observer than moralist, and his work offers neither moral corrective nor higher purpose, one might concede just the same that when Mme de Genlis tells us how well the duc d'Orléans, Louis le Gros, excelled in peasant's roles, Carmontelle's inner thought was "Indeed."[21]

The wisdom of knowing when to stop and not moralize cannot be mistaken for a lack of opinion or moral consciousness. His mockery and satire, albeit gentle, are underscored by serious intentions. Written under a pseudonym, his *Théâtre du prince Clenerzow* (1771) criticizes both French theater and society. Fashion (fashionable things) and society's blind adherence to its whims is a favorite subject of ridicule. In the Dramatic Proverb *Le Malentendu* (The misunderstood) of 1781, dieting is targeted. The proverb has several characters, including two nuns—one a sniveling fool who never finishes a thought, the other pious and gullible—and a stuttering quack doctor whose remedy for every malady is to gnaw on chicken bones.[22] In the proverb *L'Après-dîner* (After dinner) of 1768, hairstyle is attacked: an exchange between a newly coiffed, well-to-do husband and his wife goes something like this: "Do you like my new wig?" "You look like an ass." "It's the latest fashion." "For people without taste." Carmontelle extended his critique of fashion in an attack on Greek taste (*le goût grec*), a style instigated by the comte de Caylus, in which furniture and dress were given a Greek flare.[23]

When Carmontelle wrote a legitimate play such as *L'Abbé de plâtre* (The plaster abbot) of 1779, he again satirized a fashion—that of placing dressed-up, polychrome statures in gardens as ornaments. However, the play presented at the Comédie-Italienne was panned as a "veritable platitude." Ouch![24] There seems little doubt that Carmontelle was a man of strong conviction in matters of artistic taste and social virtue. That he may have formulated his pensées during the period that he was an active participant in the society he was satirizing suggests that he occupied an ambiguous, if not subversive, location within it.[25]

It is with this context in mind that one must approach Carmontelle's most ambitious, and arguably greatest creation, the garden at Monceau. It was built between 1773 and 1779, the latter year seeing the publication of the folio *Jardin de Monceau, près de Paris*. A prospectus soliciting subscribers for the folio was published shortly before the work's appearance.

With its broadsheet size, provocative text, and richly detailed illustrations, Carmontelle's *Monceau* sets itself apart. In a decade when theoretical texts on Picturesque gardens were being published almost on a yearly basis, none of the others was folio sized or illustrated; and—most noteworthy—none contained such a slight text: a mere ten-page, discursive essay that contrasted with the voluminous texts of contemporaneous garden-theory books.

As a garden Monceau also sets itself apart. It was a commission rather than the product of an owner-designer, such as Claude-Henri Watelet's Moulin Joli or the marquis de Girardin's Ermenonville. To a certain extent these owner-designer gardens were proving grounds for the garden theories presented in their respective works: *Essay on Gardens* (*Essai sur les jardins*, 1774) and *An Essay on Landscapes* (*De la composition des paysages*, 1777). As Gabriel Wick notes in this volume Monceau's owner was more interested in social status than in garden theory. Whatever political intentions the garden may have represented, it must be conceded that its owner had other ambitions. For instance Chartres's neighbor, a M. Boutin, had recently built a garden called Tivoli that was said to have been the most "magnificent" of the day. The duke was not to be outdone. In the context of the fervor for demonstrating status through building, Monceau should be recognized as an exercise in one-upmanship.[26] As such it had less to say to say about the emerging Picturesque garden theory of the time than how such gardens could be vehicles for self-aggrandizement.

When Monceau was conceived, the English, natural, or Picturesque style of garden had already been established on the continent. Yet its underlying principles had not yet been articulated. Things changed in the 1770s when several garden-theory texts were published on the new genre. In broad measure the practical and theoretical approaches of these texts could be grouped into two predominant schools, both of English origin, represented respectively in Thomas Whately's *Observations on Modern Gardening* (1770) and William Chambers's *Dissertation on Oriental Gardening* (1772). Of the two, Whately's book was by far the most influential, and the majority of French works on the subject were rooted in its theory. Chambers's book did have a following—in fact, the type of garden it espoused was very fashionable—but only one French gardening treatise of any significance could be assigned to its category: notably, Carmontelle's *Jardin de Monceau*.

The basic differences between the two approaches concerned the extent to which human agency was to be visible in the constructed landscape. How did the garden negotiate the art-nature polarity? Whately favored a more natural style, where nature, not art (read: artifice), was to predominate in the landscape created. Think of a Capability Brown landscape of infinite vistas with clumps of trees dotting acres of lawn. In a celebrated passage from his book, Whately captures the essence of the debate, pitting the "emblematical" against the "expressive." Referring to structures in a landscape—huts, bridges, statues, monuments, etc.— he called them "devices" that make "no immediate impression," because any feelings they would convey must be mediated by the mind. Rather, Whately favored landscape designs where emotions, and thus feeling, "should seem to have been suggested by the scene . . . not sought for, not labored; and have the force of a metaphor, free from the detail of an allegory."[27]

Not so for Chambers. He deemed unadorned nature insufficient to excite and please; in fact, he found nature left to herself boring. For him, a garden needed human interventions in the way of follies (ornamental structures as such), monuments, temples, and outlandish landscape concoctions to animate it: "without a little assistance from art, nature is seldom tolerable; she may be compared to certain viands, either tasteless, or unpleasant in themselves: which, nevertheless, with some seasoning, become palatable."[28]

Carmontelle was not unaware of the contemporary garden-theory debate. Whately and Chambers are cited in his *Jardin de Monceau* and prospectus, and indeed he recommends that his readers consult them for "theory," as he has no intention of entering the conversation. With his typical satirical slight-of-hand, Carmontelle removes his Monceau from contemporary discourse as if preemptively protecting himself—and his garden—from critique.

We need to remember that the prospectus and folio were written several years after construction of the garden had begun, and in reading these publications, especially the prospectus, it becomes evident that they are defenses, if not apologia, for Monceau. During the construction of Monceau, the garden quickly became subject to ridicule. That Carmontelle was stung by the criticism is evident in the prospectus. He protests that people had unjustly judged the garden to "its disadvantage," because they "couldn't be bothered to wait until the trees had grown to form masses." Taking an offensive stand, he defends his garden stating: "This sort of Garden can therefore not be judged like ordinary gardens" (prospectus, 7).

While it may have been unfair to judge Monceau before the vegetation had grown in, the extreme number of structures the garden contained was fair game for criticism. Jean-Marie Morel, a contemporary garden designer of some repute, makes a none-too-veiled attack on Monceau in his tome, *Theory of Gardens* (*Théorie des jardins*, 1776). He condemns Monceau for its hodgepodge collection of *fabriques*, lack of cohesion, and unstructured space, in which all scenes blend and conflict with each other. Morel well knew that landscapes need time to mature and develop to achieve the designer's intentions, but he was disinclined to give Carmontelle the benefit of the doubt. As a follower of Whately, Morel held a position on the nature of the Picturesque garden that was diametrically opposed to that of Carmontelle. That he so publicly criticized Monceau may have been taken as a provocation. Moreover, their personalities clashed. Morel was a humorless and dour teetotaler—a very different character than Carmontelle. They probably loathed each other. Nonetheless, Morel's criticism, coming from a well-respected and accomplished Picturesque garden designer, must have stung. Worse was the criticism of the Scottish gardener Thomas Blaikie. He agreed with Morel and did him one better: he succeeded Carmontelle at Monceau in the early 1780s and immediately began to transform the garden.

Carmontelle took revenge on Morel in a sarcastic review of his book in the *Correspondence littéraire* of 1776. The review is unabashedly partisan and summarily dismissive. Morel's important chapter on the water cycle and the central role of rain in the modeling of land, whereby the earth's complex topography allowed life to develop, was reduced to a droll précis: M. Morel's theory tells us that water runs downhill, and if the earth were flat we'd all be fish.[29]

Carmontelle's humorous rejection of Morel and the theory he espoused should not be taken as support of another theoretical approach, namely that of Chambers—who is cited in the prospectus. Rather, we must take Carmontelle at his word that his Monceau was not a model to follow and that he has no "pretention of offering . . . a theory or precepts" (*Garden at Monceau*, foreword). His endorsement of Whately's *Observations on Modern Gardening* is backhanded, in as much as his *Jardin de Monceau* is really a rejection of what the *Observations* prescribed.

The attempt to distance himself from the English tradition (of Whately) is consistent with Carmontelle's remark that Monceau was not an English garden (prospectus, 7)—a disclaimer few accepted, especially considering the strong Anglophile leanings of his patron. Carmontelle's other disclaimer—that Monceau was not a model to be followed—was equally unheeded. If anything, Monceau became the model to emulate in a society so fashion conscious that Georges-Louis Le Rouge's illustrated, multivolume publication on Picturesque gardens was subtitled "jardins à la mode."[30] Carmontelle's Monceau was just another lavish manifestation, albeit the best, of the Picturesque garden fad in France, and as such it must be approached with caution. We need only recall Carmontelle's critique of fashion to suggest what his creation of a very fashionable garden implies.

Carmontelle's Monceau exhibits the same flare for irony and satire as his writings. Indeed, just as he savaged fashion and societal trends, he subverted the new and fashionable Picturesque garden in the design of Monceau. Andrew Ayers's fluid translation of Carmontelle's *Jardin de Monceau* captures the disingenuous, if not downright odd, mood of the text, which is filled with humor, almost always aimed at social norms, fads, or practices.

For example, he invokes the desire to create a garden of "*all times & all places*" (prospectus, 7; the italics are Carmontelle's)—a trope introduced by Rousseau.[31] The philosopher's words may have been lifted, but Carmontelle brooks no sentiment for Rousseau's natural philosophy nor for the moral implications that accompany country living. He views the garden through a social lens, and it is basically hedonistic. For the

French, Carmontelle notes, the countryside is more interesting in theory than in reality: nature is damp and insect-ridden (Voilà the charms of nature!). A Frenchman can recite poetic descriptions of the country, but his appreciation of it goes little further. The "air" and "freedom" (*liberté*) of the country may be charming, but they are not really what please the French. As a gregarious people, they find their pleasure in society.

That Carmontelle should suggest the garden as a place of social intercourse is not surprising. His life was spent in observing and pleasing society, much of it at the country estates of ducal patrons. And while clean country air was considered good for both moral and physical health, the real purpose of a country estate was festive pleasure. Yet in the atmosphere of debauchery surrounding his patron and court life, the suggestion that the pleasures of society could be found in nature, where presumably the closer to nature one gets the more natural one's behavior becomes, must be considered an incitement to anarchic impulses.

Pressing his argument for the joys of country life, Carmontelle attacks the high-minded and serious intentions of the English garden. Country conversations—preferably gossip— are kept light in his pleasure ground. Philosophic ruminations are not prohibited, but please, no introspective, melancholic, solitary dreams and walks. If one must speak, by all means don't be too serious.

Carmontelle delivers a delicious critique of the "ornamental farm" (*ferme ornée*), one aspect of Picturesque gardens that many, including Watelet and Morel, had written about. It is said "that a farm can be a Garden" (*Garden at Monceau*, 16). But how so? Utility can never be united with pleasure. Society ladies can never be happy on a farm, which involves dirty work by its very nature. Furthermore, farm life is not social life and is incompatible with society's tastes and enjoyments.

He supports his argument with sarcastic commentary, using the habits and vanity of women as a basis. French society women only know rural life from paintings where it appears charming and pleasing. The reality is very different; the idyllic image one has of the farm would be destroyed if experienced firsthand.

The operative words in Carmontelle's Monceau are "to please, to amuse, to interest" (*plaire, amuser, intéresser*); they are the exigencies of a society in need of entertainment; a society that is neither capable of thinking, nor inclined to think, beyond the latest quip. It is noteworthy that Carmontelle, unlike Whately, Watelet, and Chambers, who either explicitly or implicitly place gardening in the realm of the higher arts, ignores the debate over gardening as an art form. His point is clear: for gardens to please they must descend to a popular level. The precepts of high art are not relevant. In garden art aesthetics require a quick fix: "seek out new ideas, combine them, & multiply them" (*Garden at Monceau*, 14). Considering the lofty ideals he had for art, Carmontelle's refusal to elevate the garden suggests a denial of it as high art.

If not art, then artifice. Carmontelle thus created a make-believe world filled with almost eighty different *folies*. These included rustic structures, exotic pavilions, faux ruins, sculptures, and a variety of landscape features. All was crammed into less than thirty acres for an intended two-hour *tour de dégustation* (less than two minutes per object, excluding walking time).[32] Such a design did not allow sufficient time for associative imagination and reflection.

Clearly, Carmontelle's intentions are different. Monceau was not meant to be a garden for contemplation, but rather a garden of theatrical illusion. It could be likened to a series of operatic stage sets brought forward individually for the evening's entertainment. Only, instead of stagehands changing the sets visitors to Monceau uncover the scenes as they explore the garden.

In its day Monceau was compared superficially to Stowe, because of the number of buildings and structures in each. In reality Monceau bears little resemblance to Stowe. The English garden was the product of a succession of architects and landscape designers—Bridgeman, Vanbrugh, Kent, Gibbs, and Brown among them. Built over many years, it is a designed landscape with multiple, complex layers of association and meaning, including antique, Renaissance, allegorical, and political references. Consider its temples—Ancient Virtues, British Worthies, Modern Virtue. Now consider Monceau's White Marble Temple—literally a white, monopteral, marble temple (plate XV). An allegory or a joke? Carmontelle imputes a story to the temple. Achilles is recalled: "Placed on the altar, which is in the middle, is an antique figure representing one of Achilles's female companions when he was at the court of King Lycomedes" (*Garden at Monceau*, 22). Misreading Carmontelle can be treacherous. Here we have Achilles, the Greek tragic hero, the prototype of manly valor and beauty. This, however, is not the aspect of the hero's legend Carmontelle cites. Rather, Carmontelle focuses on an uncharacteristic chapter of Achilles's biography: to escape the mobilization of the Trojan War, Achilles hid among the members of the court of King Lycomedes on the island of Scyros, dressed as a woman. While there he seduced Lycomedes's daughter, Deidamia; Neoptolemus was their issue. The legend continues: Ulysses, in search for Achilles, arrives at Lycomedes's court disguised as a merchant. As he shows his wares, a cache of arms falls from his sack and Achilles quickly grabs for them.[33]

The reference to Achilles's distaff experience and masculine recovery would not have gone unnoticed. Exactly the opposite: the point of society was to know enough to be in on the joke and move on. That the legend would have had resonance with the society crowd of the era is suggested by the life of the chevalier d'Eon. D'Eon was a decorated military officer and confidant of Louis XV. When stationed in London, he began to dress, if not live, as a woman. He returned to France in 1776, expecting to continue to live as he had in London. The confounding, not to mention outrageous, story of the chevalier/chevalière d'Eon is precisely the kind of story Carmontelle would have exploited at Monceau.[34]

For those not versed in mythology, Carmontelle offered the Horse Chestnut Chamber (Salle des Marronniers, plate XVIII). The name derives from horse chestnut trees that encircle an elevated platform; the platform has a niche containing a statue. Carmontelle's description is ludicrously precise, down to the slightest, easily missed detail. What is not easily missed is the contemporary copy by Edme Bouchardon of the Barberini Faun—the original is probably Hellenistic—whose debauched exhaustion is prominent. Lest anyone miss the point, Carmontelle depicts a garden guide, accompanying three women, who points to the groin of the sated satyr. The body language of the women suggests that they understand the intent of the guide's gesture.[35]

Carmontelle's sexual playfulness is also found in the Chinese brass-ring carousel (plate XVI). Women sit on pseudo-Chinese male mannequins whose outstretched arms disappear provocatively under the dresses of their female riders, while men sit astride dragons whose necks rise phallically in pursuit of the women.

Perhaps the wittiest, and easily the most ambiguous, feature of Monceau is the Naumachia and its primary feeder stream. Seen in plan, it suggests the form of a sperm cell, right down to the tapering, serpentine stream-tail and the flattened oval head of the water basin (plate 1). The conjunction of oval basin and stream is too specific to be coincidence. Of all the water features in the gardens illustrated in Le Rouge, none similarly reproduces the morphologic exactness of Monceau's sperm cell.[36]

The introduction of sperm into Monceau is outrageous but not inconsistent with the garden's program of pleasure, nor the consciousness of the time. The nature of sexual generation had been an issue of fascination since antiquity. Sperm went unobserved until Antoni van Leeuwenhoek published his "Observationes de natis e semine genitalis animalculis" in the *Philosophical Transactions of the Royal Society* (London, 1677). At the same time Christian Huygens, working in France, was also investigating the nature of the male seed. In the following century, Georges-Louis Leclerc, comte de Buffon, the *grand homme* of French natural history, published his findings on generation in his *Histoire naturelle*.[37] He examined and illustrated the sperm of dogs, rabbits, rams, and fish, and reproduced Leeuwenhoek's drawings as well (fig. 53). Indeed, Monceau's "sperm cell" bears resemblance to that illustrated by Buffon.

More politically pointed, the symbolic spilling of the male seed may have been a not too veiled commentary on the supposed infertility of Louis XVI who had yet to produce a child, let alone a male heir. At the same time it oculd have served as a statement on the apparent prodigality of the duc de Chartres.

Sperm, sexual generation, and science were not only games of a bored elite, but were part of a more serious concern of eighteenth-century France: Freemasonry. When the Grand Orient, the comte de Clermont, died in 1771, French Masonry was on the verge of anarchy. No doubt for political reasons the young duc de Chartres was asked to take command. He accepted and established his lodge, Saint-Jean de Chartres, Orient de Mousseaux, on December 20, 1773.[38] It survived until 1789.[39] The duke brought the prestige of a *prince du sang*, but he was far too unstable for leadership. Thus his Masonic responsibilities were more honorary than administrative. Not inclined to focus, he was said to bore easily when forced to exercise his duties.[40] At his private lodge at Monceau, he oversaw his own brand of Masonry—one more social than cabalistic.[41]

Carmontelle was not a Mason, but there is no doubt that he was aware of Masonry. He was probably amused by it and dealt with it as he did the other elements of his garden, using the well-known social trope of persiflage to *persife* or trivialize and mock its pretenses. In his aforementioned proverb, *Le Malentendu*, he lets us in on the joke. One of the characters, le Père Saturnin, whose name is a Masonic pun, mouths Masonic prescriptions: "With philosophy we know about action and reaction; the atmosphere, properties of air, water, earth, and fire."[42] Father Saturnin is the Masonic high priest, but inasmuch as the proverb is only a few pages long Carmontelle remains more interested in the implied pun than in its development as a broader narrative. Thus the association of elements of the garden at Monceau with Freemasonry remains in the category of a conundrum.

❧ ❦

I N Carmontelle's Monceau the Masonic pyramid becomes old Ninny's tomb as boundaries of meaning slide from ridiculous to sublime. Nowhere is this more evident than at the Wood of Tombs (plate XII), which offers, we are told, a fitting setting for Pyramus and Thisbe, the classical myth popularized by Rebel and Francoeur's opera, produced in 1726 with set designs by Giovanni Nicolò Servandoni. No doubt as a man of the theater, Carmontelle knew Servandoni's work, if not the man himself, and could appreciate— or envy—the Italian's superior talent. It is probable that he also had personal contact with the opera. He could have seen Sophie Arnould, who sang the role of Thisbe for the first time on January 23, 1759, and performed it through the spring of 1771. By then she had achieved prima diva *stupenda* status, as well as a reputation for being a woman of loose morals.[43] It was then, too, that Carmontelle drew her portrait in the role of Thisbe "in the middle of a dark forest filled with tombs"— including a pyramid. The joke at the tomb is not diva Arnould's character or even her voice, which by 1771 was failing.[44] Rather Carmontelle is recalling Shakespeare's comic interlude in *A Midsummer Night's Dream*—the savage parody of Ovid's Pyramus and Thisbe legend.

Carmontelle was surely familiar with *A Midsummer Night's Dream*. The play's enchanted fairy kingdom, juxtaposed to that of a human world of reality, not to mention its famous satirical episode, must have appealed to him. Puck, Quince, Snug, and Bottom could easily have inhabited the garden at Monceau. That Carmontelle did know the play and its frivolous aspects, particularly the Pyramus and Thisbe episode, is suggested by his dramatic proverb *Le Jardin anglais* (The English garden), which includes a sly reference to Pyramus's plea to his beloved: "Wilt thou at Ninny's tomb meet me straightway?"[45] The proverb is typical of Carmontelle: short, dull, silly, and filled with sentimental amorous intentions and polite sexual innuendo.[46] The dramatic action turns on the tragic circumstances of Pyramus and Thisbe, except that in

(150)

Carmontelle's *jardin anglais* the lovers don't die. The marquise's lover, the chevalier, has been away in England for six months. In his absence she has transformed her formal French garden into a *jardin anglais* equipped with a windmill, a grotto, a large lawn, a ruin, and a tomb. During their separation they have had little communication; each is anxious about the other's well-being. Wishing to surprise his mistress, the chevalier returns unexpectedly. On the day of his return, the marquise is overseeing work at the tomb. The chevalier enters the marquise's property by an alternate route; he comes across a servant boy who is crying because his father has just spanked him for insubordination. The chevalier—noting the youth's distress— asks straight away after madame. "Madame la marquise, monsieur?" the boy replies. "Oui!" the chevalier barks nervously. "Eh bien, she is at the tomb." The chevalier collapses in tears. He is revived in the arms of his love amidst a fantasy of an English garden. There isn't any more to it.

Invoking Shakespeare's Pyramus and Thisbe is vintage Carmontelle. He dropped satire into the heart of his garden. In fact, throughout he had eschewed all serious pretensions. Here we have Achilles in drag, semen-stained earth, men with arms under women's dresses, and old Ninny's tomb: are these the things that move men's souls? Carmontelle's pleasure garden is not blatantly obscene; it's more like the Marx Brothers at the opera, a landscape of entertainment that aspires to no other purpose than pleasure. But at the same time Monceau burlesques the English garden, it ridicules equally the members of society who inhabit it. Carmontelle was society's wizard, replacing each evaporating whim with something more delightful to those engaged in an endless pursuit of idle pleasure—pleasure based not on art, but on immediate gratification. As a physical counterpart of the Dramatic Proverb, Monceau offered no higher principle. A product of fashion, Monceau soon became its victim, and within a few years of the garden's completion, both Carmontelle and Monceau were no longer in vogue.

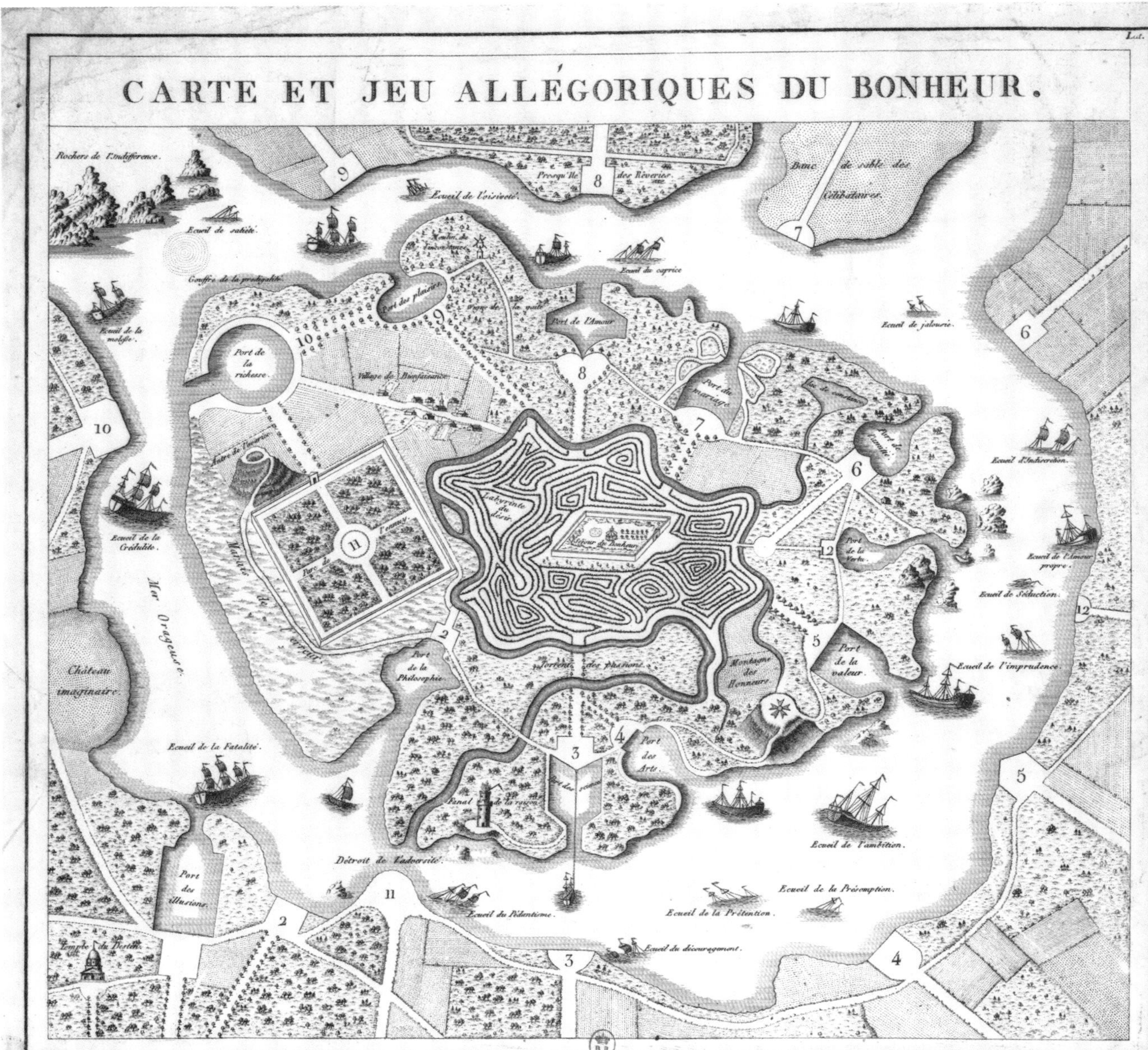

FIG. 54 *Carte et jeu allégorique du bonheur*, ca. 1780–1800.

EN-JEUX: VIEWING, MAPPING, AND PLAYING IN CARMONTELLE'S PROSPECTUS AND *JARDIN DE MONCEAU*

Susan Taylor-Leduc

IN HIS FOLIO *Jardin de Monceau* and the prospectus that preceded it, Carmontelle made it clear that his garden at Monceau was not to be considered an English garden. But perhaps realizing that his anti-theory might be confusing, he included in the folio a plan keyed to the engraved views, with written descriptions of the engravings in order to help potential visitors and readers better appreciate the garden. The engravings continue to enchant us today. Figures gesture towards miniaturized buildings or pause to admire masses of shrubbery and flowers where cows and sheep mingle with exotically costumed figures. As the prospectus announced, the engravings were easily procured, whether as a bound volume or separate sheets. Sold at a range of prices, beginning at 30 sols for a single print, they targeted a rather elite audience. Subscribers could purchase the total suite (text, plan, and seventeen views) for an upfront investment of 27 livres. Afterward the folio would be available for 40 livres at booksellers, engravers, and *marchand-libraires*.[1]

We can speculate about how reader-viewers may have interacted with Carmontelle's luxury edition. For example, the folio could have been read from the comforts of their salons where they could coordinate plan, descriptions, and engraved views, perhaps in anticipation of their visit to the latest *folie* of the duc de Chartres.[2] Alternatively, despite the fact that a folio volume was a bit unwieldy to carry, visitors could bring the work to the garden to serve as a guidebook.[3] It is also possible that visitors purchased the engravings as souvenirs after their visits to the site.[4] For those who could not visit, the folio was a graphic delight that evoked an imaginary tour of a fashionable paradise. For all readers Carmontelle's luxury edition provided unprecedented views of the garden at Monceau that were meant to incite curiosity.

Scholars have acclaimed the ten-page text, the plan, and the seventeen engraved views as major contributions to eighteenth-century French garden theory.[5] This essay does not address Carmontelle's position as a theoretician—or anti-theoretician—per se but rather specifically focuses on the relationship between the descriptions and the engravings. Carmontelle divided the text into three chapters that contain descriptions of the engraved plates along with some theoretical observations (the latter occupying the first few pages of chapter 1).[6] I argue that in these chapters Carmontelle devised a subtext as an ancillary reading that complemented the more theoretical introductory pages and the prospectus. Although Carmontelle claimed that the descriptions and engraved plates would elucidate his theory, a closer analysis reveals that the chapters elicited confusion rather than clarification. They provide directions that are deliberately befuddling and do not necessarily explicate the scenes depicted in the plates. The disparities between reading, viewing, and strolling suggest that Carmontelle treated the folio edition as an artfully designed game that mimicked the playful encounters he created for the garden. Considering the folio and garden as an interconnected game-scape offers an opportunity to reevaluate how we interpret the art of strolling at Picturesque sites.[7] Further, acknowledging the ludic as an essential aspect of Picturesque design enhances our appreciation of how

Picturesque gamescapes metamorphosed into *jardin-spectacles* during the Revolutionary decade.

Before turning in detail to Carmontelle's text and plates, it is helpful to review the market for prints and illustrated luxury folios in the late-eighteenth-century Paris in order to better understand how Carmontelle deployed the edition as a foil to both his theory and his garden's design. As literacy increased throughout the century, collecting prints and books with engravings appealed to elite connoisseurs and an increasingly bourgeois social milieu.[8] Prints were available for purchase from publishers of illustrated books, print merchants, and—increasingly after 1770—at regularly scheduled auctions.[9] Whether savvy customers acquired folio editions or separate prints, they helped to drive an expanding market for a new repertory of images including landscapes, religious pictures, secular subjects, and libertine genre scenes.[10] Prestige editions appealed to connoisseurs, garden theorists, and critics who appreciated the scenic qualities of the prints, but also, as Carmontelle realized, to a Parisian fashion and consumer culture that contained diverse audiences seeking amusing distractions.

The plates depicting the garden at Monceau reflect an ongoing admiration for engraved maps and topographical views. As David L. Hays reminds us, plans of Picturesque gardens that used some of the techniques of surveying and mapmaking appealed to print connoisseurs and garden amateurs alike.[11] Georges-Louis Le Rouge's publication of over four hundred garden views in his *Jardins anglo-chinois,* issued in approximately twenty loosely bound cahiers from 1775 to 1789, attests to the popularity of both maps and garden plans. Le Rouge, a surveyor and enterprising print publisher, slightly modified several of the Monceau prints and included them in his tenth cahier.[12] It appeared in May or June 1783, suggesting the increasing popularity of garden imagery in the 1780s.[13]

Given the expanding appeal of prints in the last three decades of the ancien régime, we can hypothesize that Carmontelle's plates were not only valued as illustrations of the garden at Monceau but were also appreciated as autonomous visual artifacts that revealed a wealth of information about contemporary mores.[14] For example, the figures that populate the engravings are costumed in the latest dresses and shoes for strolling and display various accoutrements such as fans, parasols, and walking sticks—as discussed in greater detail by Caroline Weber in her essay in this volume.[15] The women model the latest hairstyles, which include hats, feathers, and ribbons popularized by fashion merchants. Men sport trendy costumes as well, with hats and carrying walking sticks. Notably, two male strollers hold either telescopes or Claude glasses—curved lenses that enhanced the viewing experience so that the scene had an augmented resemblance to landscape painting. Some of the figures are dressed in exotic costumes, testifying to the access the duc de Chartres and his couriers had to foreign markets and to imported goods for sale in Paris. The range of buildings, from minarets to ruined temples not only reveals the dukes unlimited funding for his garden but also his desire to display the most stylish decorations. The structures recall examples seen on popular printed sheets of architectural ornaments. The landscape itself features densely planted shrubberies and flower beds stocked with plants that were available for purchase from international nurserymen.

The plates confirm that strolling at Monceau was a popular elite leisure activity and an opportunity for self-fashioning.[16] The two pastimes, strolling and shopping, were intimately connected.[17] By the 1770s promenades on the boulevards atop the former ramparts of the city provided places to stroll and shop, a combination of activities that would be available on an expanded scale at the Palais-Royal.

When the duc de Chartes received the title of duc d'Orléans upon the death of his father in 1785, he began to transform his family's Parisian residence and formal garden at the Palais Royal into a shopping emporium.[18] It rivaled the traditional fairs and such pleasure gardens as Vauxhalls, offering amusements all year long.[19] The creation of arcades around the formal garden attracted a range of merchants, who competed for customers' attention with street theaters whose performers staged marionette and shadow shows (*ombres chinoises*) and mingled with charlatans and magicians enacting prestigious feats (often abetted by new technical devices).[20]

Carmontelle was intimately familiar with the French entertainment industries. As general director for entertainment in the Orléans household, he was expected to not only follow but also to initiate trends, to amuse his patrons, assure they were never bored, and enhance their reputations as models of fashion. Given that Carmontelle was a recognized playwright, art critic, and party planner, it seems predictable that he would turn to theater for inspiration in the design of his garden and his prints. A sequential reading of the engraved suite was explicitly scenographic; the arrangement encouraged reader-viewers to imagine the garden as a series of unfolding stage sets. Turning the pages, they witnessed an ongoing spectacle that conjured up imaginary tours. This scenographic organization of the garden reflected Carmontelle's own expertise; however, theatricality was one reference amongst many that informed his project.

Carmontelle stated explicitly that he conceived of the garden at Monceau as "pure entertainment"— wording that stressed playfulness as a primary inspiration for the design. He explained that he needed to fix the attention of his patron as well as attract a wider public that expected to be amused. Accordingly, he devised a detailed system for calibrating plan, description, and engraved views. Each plate was numbered, with the numbers keyed to the descriptions in the text. The vantage point stipulated for viewing each scene was identified by a letter that appeared in its description and on the plan. This elaborate system suggests that the descriptions and plates reveal information about the garden, yet the subtext has a decidedly comic twist.

Plate V, *Of the Farm, taken from point D, near the Cabaret*, will serve as a case study, allowing us to examine how Carmontelle constructed an explicitly playful connection between the folio and the garden. The designated vantage point (D) is easy to locate on the plan, on the northern edge of the plate near the rue de Monceau. In the view Carmontelle depicts vernacular architecture in the background and places two cows strategically in the foreground. It is difficult to discern a specific building on the plan that could be the cabaret referenced in the title. Two costumed male figures are placed in the left foreground of the view; their dress is perhaps Turkish, perhaps Amerindian, but they are certainly non-Western. The turbaned figure leans on a staff; possibly he is the cowherd. By contrast, a group of four stylishly dressed figures is found to the right of the animals: one man and two women are conversing; the fourth, a woman seen in profile, turns her back to the cows and approaches the group on the right.

When we read the description of plate V, we begin to realize that Carmontelle confounds rather than clarifies meanings. His description and the view are disconnected rather than clearly contingent upon one another:

> One sees the Buildings on the left, of which one can make out only the top of the Gardener's dwelling & the cow sheds; & those in the foreground are the goat & sheep sheds. At the rear is the dovecote behind which are the hothouses, the steam pump, & the fig orchard; & in front, the wood containing the little rock fountain that was mentioned in the description of Plate III.
>
> Turning around, one can take the path to the right, running alongside the Monceau road, & one will arrive near point E. (*Garden at Monceau*, 18–19)

From this we learn more about the buildings, which ostensibly house animals—they seem rather small to support farm labor in this suburban space—and we discover that there is a fountain covered with shrubs, yet we gain no additional information about the figural groups. We do have an enumeration of areas that are *not* visible in the engraving: the hothouses, the steam pump, and the fig trees.

Carmontelle suggests these areas were discussed in plate III. Quickly turning back to plate III, *Of the entrance to the Garden, taken from point* B, *on the Isle of Sheep*, the reader finds that Carmontelle does not illustrate the hothouses or fig orchard. Moreover, the description of plate III is cryptic and primarily offers more directions:

> After crossing over to the Isle of Sheep, at point B, one can see the entrance to the Garden, which is next to a small Gothic Building that serves as a Chemistry laboratory & has a Turret that terminates in a point. Under the arcade of a crenelated wall the river passes. To the right is the court that is in front of the main pavilion, in the middle of which are several trees, grouped around a perch, which is intended for shooting the popinjay . . .
>
> On leaving the Isle of Sheep, to the right, there is a little Fountain set into a rock on top of which there is a ruin. If one passes in front of the Mill bridge & under the drawbridge gateway, & one then climbs atop the Mill, by turning to left & right one can get a good idea of this part of the Garden.
>
> Going down to the right, crossing the Willow bridge & again going right, one will find a sunken path lined with woody nightshade made to provide shelter when there is too much sun. After passing under an aqueduct one will find oneself, having turned left, at point C. (*Garden at Monceau*, 18)

By now the reader-viewers are completely confused: neither of the descriptions seems to fully elucidate the views, and the texts refer to areas that are not depicted on the plates.

At this point, they turn to the plan, hoping that it will provide more information. There point D is clearly aligned towards the Farm, and the location of the hothouses becomes clear: they are located behind the Farm, but are not visible in the engraved plate. Rather, Carmontelle closes the view beyond the Farm with a row of trees that corresponds to similar rows used on both plan and engraving to hide the fig orchard and hothouses. Consequently, what we are supposed to see is, in fact, masked. Considering the descriptions and plan together helps to elucidate the orientation of the view, but the engraving does not offer us a better understanding of the itinerary or the spatial alignments of the garden. We are instead immersed in a world of visual play, where moving from one scene to the next becomes a fanciful endgame disenfranchised from the actual garden.

Returning to the plan, we can look at three vantage points and articulate the relationship between the stroll and the view: From point D to point E is a rather straight line, either on the path or across a lawn to the edge of the garden. Yet from point B to point D the route is confusing: either we follow the serpentine path in the woods, arrive at the small fountain mentioned in the descriptions, and then continue to point D by following the path bordered with woody nightshade (*vigne de Judée*) as noted on the plan, *or* we follow Carmontelle's instructions: cross the water path, traverse the Isle of Sheep, visit the Mill, cross the Willow Bridge, and then turn right onto the nightshade-bordered path, which passes through a shady area and under an aqueduct to reach point C. Unless the visitor had the descriptions in hand, it is difficult to know how to follow the complicated instructions Carmontelle provides. Moreover, it is equally problematic to imagine the same visitors seen in the plate V, dressed in silken walking slippers (mules) and strolling costumes, climbing atop the mill to have a view from above to see over the garden, yet this it is exactly what Carmontelle suggests.

How do we make sense of these befuddling descriptions? Carmontelle provided his first hint that the engraving of the Farm was not meant to be representational in his first chapter when he explained that he did not expect the Parisian crowd or members of the Orléans family to appreciate farming:

> We would not be able to enjoy the cares of a farm; its multiple occupations leave no time for leisure, & the details of rural life are ill-suited to our taste for society, pleasure, & dissipation; we prefer the description of them to the practice. Do we think that most of our Ladies, who sit in a Drawing

Room whose shutters are tightly closed when the sun shines, would find it charming to follow the labors of a farm the whole day through, to oversee the harvests & the grape gatherings, & to watch the tending of the animals? (*Garden at Monceau*, 15)

Although Carmontelle included sheep overseen by a turbaned shepherd on the Isle of Sheep (plate II) and a lone cow waiting to be milked in plate XV, these farm animals are ignored, notably by the ladies. By contrasting cosmopolitan and foreign figures with the cows and sheep, all the personages are removed from any real engagement with farming or pastoral reverie; rather they seem to pass through the scene en route to the next amusement, leaving the farm animals behind. Carmontelle thus champions a double entendre: he allegedly records a scene in the garden, but the composition mocks the subject. As Joseph Disponzio and David L. Hays have noted, the figures undermine the very precepts of contemporary Picturesque garden theory.[21] At Monceau the pastoral tradition is therefore exposed as farce for both visitors and readers.

Similarly, the presence of foreign figures in plate V suggests several parodies. Carmontelle explained that "foreigners" in his views were fake: "The foreign figures who are in these drawings are the persons who serve the Prince in different costumes when he dines or sups in his pavilion, or strolls in his Garden" (*Garden at Monceau*, 18). This statement asserts the princely status of the duc de Chartres—he can hire whomever he wishes and have them dress as he pleases—yet, as in the parody of the ornamental farm (*ferme ornée*), Carmontelle seems to undermine the Picturesque itself. The exotically garbed figures may reference William Chambers's *Dissertation on Oriental Gardening*, which ostensibly linked Monceau to the international taste for orientalism, but as servants dressed in exotic costumes, they turn the issue of the validity of importing foreign styles to France into a satiric commentary rather than representing a sustained interest in the Far East.[22] Nonetheless, the representation of non-Western figures was intentionally surprising: they emblematize for the reader-viewers-strollers the fact that they would encounter characters whose presence was unexpected. They promised that Monceau was interesting precisely because it presented *"all times & all places"* (prospectus, 7): an accumulation of spellbinding entertainments and luxury goods.

The subtext of plate V is a critique of Picturesque theory. However, the connection between the plate and its description becomes more coherent if we consider that the linkage of the points on the engraving to the plan functioned as an elaborate game. It is possible that by defying narrative legibility and creating a deliberate disjunction between image and text, Carmontelle was visualizing a joke.[23] Perhaps for reader-viewers the plan and engravings functioned as enigmas, a literary device that was often employed to ridicule.[24] Enigmas were well entrenched as a form of courtly amusement, but for the Orléans courtiers, who were not intellectually inclined, the most significant form of pleasure was not in decoding riddles but in gambling.

Gaming was ubiquitous in eighteenth-century Paris: the maître de Claville, in his etiquette guide, *Traité du vrai mérite de l'homme*, first published in 1734 and reissued in 1761, explained that knowing how to play parlor games was an essential skill for polite sociability.[25] Yet for the Orléans entourage—especially Chartres, who was known to be an obsessive gambler—high-stakes card games and betting occurred daily.[26] Given the omnipresence of gaming, reader-viewers brought their own familiarity with the practice to their appreciation of the plan in the folio. The art of gambling could be transposed onto the garden experience. Gamblers who became strollers did not abandon their gaming skills at the garden gate; rather, each "chance encounter" while strolling incarnated the psychosocial thrills experienced when a winning card was revealed or the roll of the dice delivered a triumphant victory.

Illustrated books, prints of board games, and decks of cards were readily available at boutiques run by print sellers (*marchands d'estampes*) and bookstores. There was a specialized print market dedicated to gaming manuals and rule books that probed mathematical probabilities to suggest winning strategies to help players avoid being duped.[27] Books were updated when new games, such as whist, appeared and even when almanacs recorded the latest gaming trends.[28] Carmontelle, whose career was predicated on finding clever and pleasing means to entertain the Orléans court, certainly understood the ability of both board and cards games to surprise and amuse, captivate, and most significantly, provide pleasures. Moreover, for gamblers gathered at the Palais Royal—meeting in private clubs, wagering on street games while standing under the garden arcades, or gaming behind closed doors of the Orléans salons, it often resembled a *tripot*, or gambling house.[29]

I am not suggesting that the garden at Monceau is to be considered a reconstitution of a particular game; rather, I contend that the culture of gambling offered new paradigms about space, mobility, and emotional engagement that would have appealed to the risk-oriented social milieu that Carmontelle was expected to entertain. We can distinguish how gambling culture could be transposed to garden design by examining the Monceau plan. As Disponzio has pointed out, the plan "exhibits no sequencing of spaces, and no formal or axial relationships . . . The design is loosely structured, without focuses and devoid of linkages among its parts."[30] Hays has observed that the descriptions and plan reveal a horseshoe-shaped itinerary, where at several junctures it was possible to choose a path to discover different areas on the approximately thirty-acre site. The plan is, at the very least, confusing, and denies a clear linear path or single itinerary. Yet when it is considered as a board game, we can begin to unravel how the plan and engravings interacted with the physical experience of strolling.

Michel Conan has addressed how *cartes de tendre*, fictive maps dedicated to ideal friendships, informed seventeenth-century French garden design, providing a model for courtly behaviors when people were promenading.[31] By the later eighteenth century, the moralizing emphasis of the *carte de tendre* had fallen out of fashion. Instead, as the board game called the Map and Allegory of Happiness (*Carte et jeu allégorique du bonheur*) demonstrates, chance replaced literary allusions and verbal badinage.[32] According to the rules of this game, players roll the dice to determine their movements from the mainland to a central island. The goal of the game is to arrive at the miniature labyrinth in the center, identified as a space dedicated to happiness (*le séjour de bonheur*), after having traversed emotional states such as rocks of indifference (*rochers des indifference*), sand traps for the single person (*banc de sables des célibataires*), the grotto of greed (*antre d'avarice*), and the cascades of passion (*torrent des passions*) (fig. 54).

Comparing the plan of Monceau to the printed Map and Allegory of Happiness suggests how board games would have appealed to Carmontelle as paradigms for spatial design. On board games players advanced with each throw of the dice. When Carmontelle designed paths for his strollers, he likewise privileged chance encounters over prearranged itineraries. The swirling paths and strategically placed obstacles recall how pure accident determines how one advances in board games. For Carmontelle referencing the board game favored happenstance, whereby surprise became a fashionable means of animating a stroll.

Another board game, the game of Goose (*Jeu de l'oie*), further elucidates how gaming could be transposed to garden design. Its plan consisted of a spiral of sixty-three squares. Players adopted pieces (often small geese) that they advanced around the spiral by throwing dice until they arrived at the center. Chance determined if one lost a turn, returned to the beginning of the game, or accelerated to the end (fig. 55). Progression and regression on the board provided another model for strolling. It suggested that strolling was not limited to a forward-moving linear sequence, as in the French allée; instead one could advance, pause, and return, depending on the unpredictable roll of the dice. In his descriptions, Carmontelle advises his strollers that they should twist, turn, and change their paths in order to reach the next stop, recalling the motion of board games.

At both Chantilly and Choisy-le-Roi, the game of Goose replaced the geometric path of labyrinths (fig. 56). On a contemporary plan of Choisy, the game was confined to a bosquet, suggesting that the royal gardeners were integrating a new format into an existing structure. Le Rouge published an engraving of Choisy in the same cahier of his *Jardins anglo-chinois* as his Monceau prints, presenting the *Jeu de l'oie* as a design motif that rivaled the traditional maze. Carmontelle may not have visited Chantilly or Choisy, but we can imagine that he was familiar with the game of Goose as a form of entertainment and a new paradigm for garden paths. Likewise he probably knew of the game's pedagogical application in aiding memorization, a tool that would have helped in the education of his pupil, the duc de Chartres who was an indifferent learner. Certainly Carmontelle would have remained sensitive to educational concerns in the Orléans household, given that Chartres's mistress, the comtesse Stéphanie-Félicité de Genlis, who officially educated the duke's children and would later publish books dedicated to pedagogy, shared status with him as a privileged employee at the center of Orléans entertainment culture. As a form of flattery and complicity, we can imagine both Carmontelle and Genlis appreciating the *détournement* of the game of Goose from its pedagogical function to adult entertainment in order to please their mutual patron.[33]

Card play provided further examples of how the social praxis of gambling could be integrated into garden strolling. Cards were omnipresent in eighteenth-century Paris: they were widely recognized to connote chance, unpredictability, and surprise. Card play was divided into two types of games: either *jeux de levé*, where the winner was determined according to the card revealed at the end of a round of play, or combination games in which players selected cards from a deck and then strategically organized them to constitute a winning hand. Most card games used both modalities. All the card games were contingent upon luck and anticipating the end of play (who would have the winning hand?); expectancy of the unpredictable was mesmerizing entertainment. Gaming was considered a happy and distracting amusement during which one awaited surprises.[34]

The transfer of the emotional engagement of card play from the gambling table to the garden materialized when strollers expected, pursued, and performed surprise. Contemporary visitors were encouraged to discover eighty miniature constructions (*fabriques*) in the garden; the space was jam-packed to elicit wonder. In addition, Carmontelle's hired servants disguised as hermits or Turks were deployed to intentionally startle visitors. In contrast to the corporeal restrictions on bodily reactions to surprise at the card table where one was expected to deploy a series of graceful gestures to hide one's hand while hedging a bet, performing acts of surprise in the garden was freer from physical restraints.[35] While strolling, one could indulge in the pleasure of being surprised and relish bodily frisson and emotional glee as affirmative physical sensations. This desire to pursue surprise in gardens was often linked to the allegorical pursuit of love and carnal engagement, supporting Disponzio's contention that Monceau was a space that promoted libertinage and sexual profligacy.

Shortly after publication of the *Garden at Monceau*, Chartres began offering tickets to visit the garden, suggesting that it was recognized as a place to seek entertainment. The construction of the carousel (*jeu de bague*) recorded on plate XVI introduced a physical game as one of the major attractions in the garden. Carmontelle explained how the carousel worked in his description of it, while at the same time alluding to its sensationalism as a game:

> The carousel is a Chinese parasol held up by three Chinese figures, which also hold a horizontal bar that is used by those who turn the carousel, & who need do nothing other than walk on the wooden boards beneath their feet. From the edge of the boards, four iron branches extend, of which two support dragons that can be mounted like a horse. At the end of the other two branches there are re-

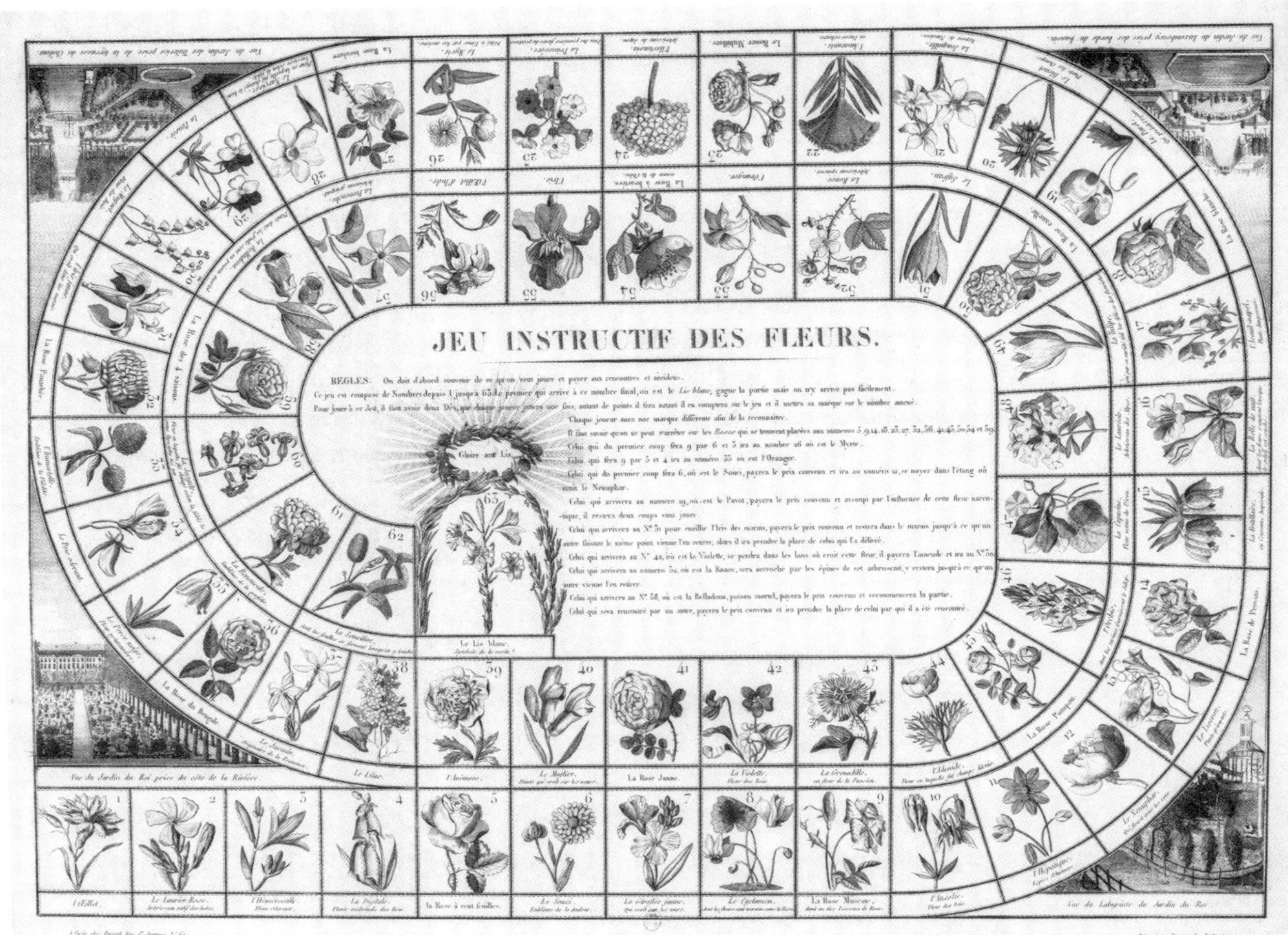

FIG. 55 *Jeu de l'oie, Jeu instructif des fleurs*, 18th century.

clining Chinese costumed mannequins; with one arm they support a cushion on which one may sit, & each holds in his hand a parasol trimmed with little bells; with the other hand they hold a cushion for the feet. Ladies sit on these two branches.

The edge of the large parasol is trimmed with ostrich eggs & bells. The four lanterns that can be seen contain rings at the end of the bobbles beneath the lanterns that are only visible to those riding the carousel. (*Garden at Monceau*, 23)

For reader-viewers, libertine allusions transformed the scene into one of orientalist fantasy. The snakelike seats, the disguised men who literally support the women by reaching under their dresses, and the overall orchestration of bouncing movement demonstrated that the garden was a place dedicated to sensuality, sexuality, and physical delights. Carmontelle outfitted a bench as an ottoman under a silken tent so that visitors could watch the swirling fun. Trendsetting visitors invited to take a turn on the merry-go-round experienced vertigo: a dizzying appreciation of color and forms.[36] The garden had become a playground; a space dedicated to moral emancipation.

In the prospectus Carmontelle wrote that "one wishes to spend only a few hours" and estimated that "it will take nearly two hours to walk through [the garden] without stopping" (prospectus, 8). He added: "If the main features were further away from each other, one would have to visit it on horseback or in a carriage. Had we included fewer features, there would have been less variety, & the different scenes that appear

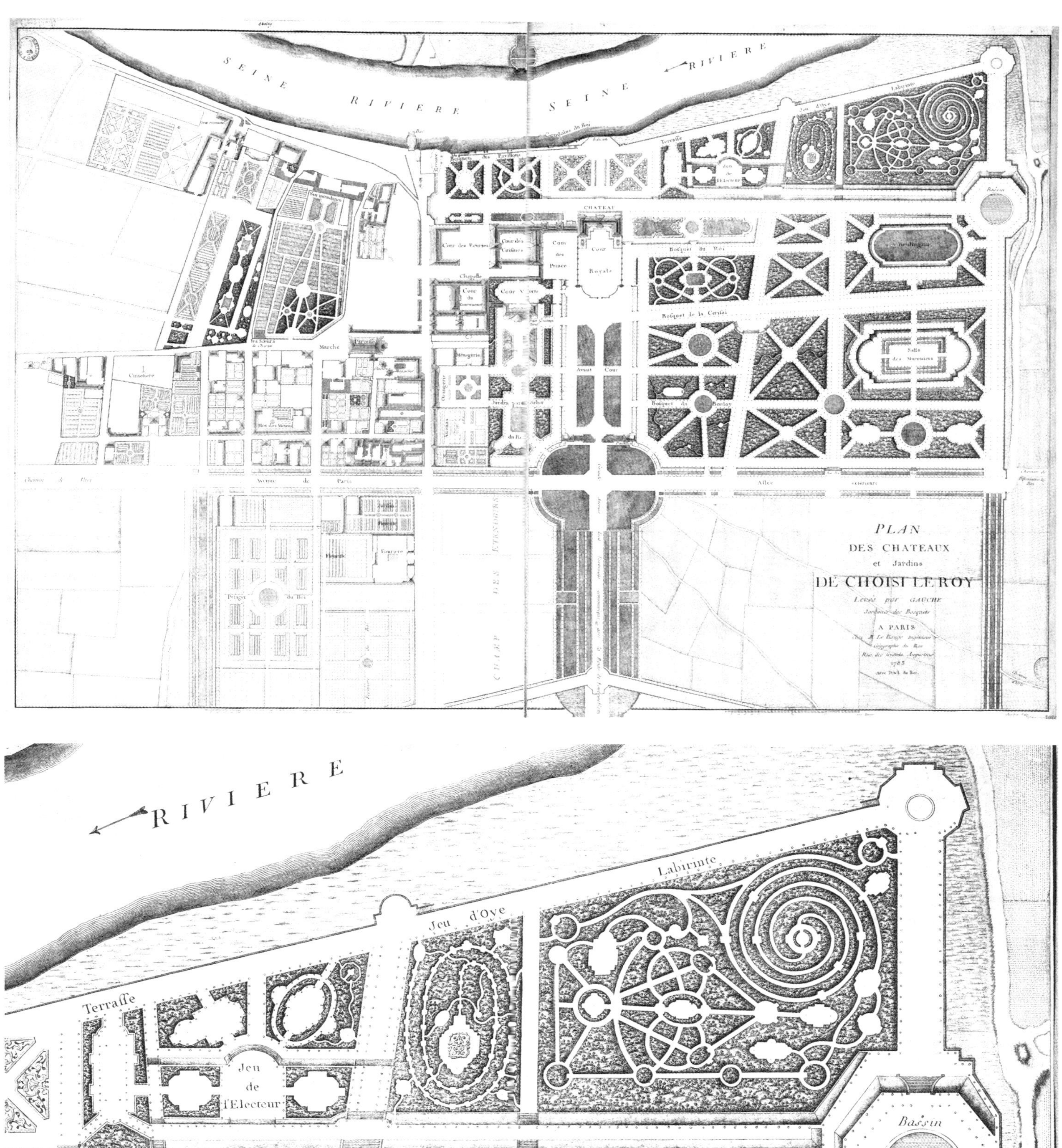

FIG. 56 Plan and detail of the châteaux and gardens at Choisy, 1783, published in G.-L. Le Rouge's *Jardins anglo-chinois*.

at each moment are the source of all the pleasure" (prospectus, 8). Thus the stroll in the garden was supposed to be a brisk but immersive activity. Visitors played on the carousel or were constantly on the move, discovering a landscape transformed into a gamescape that enhanced the ephemeral pleasures of surprise and ensured that they were never bored.

The notion of the garden as a gamescape in the decades preceding the fall of the ancien régime echoes the economic, social, and political anxieties of a period when the social order itself was in play, a subject beyond the purview of this essay. For Carmontelle, however, play was a useful guise: he teasingly subverted the relationship between text and image, denying legibility in favor of entertainment. For reader-viewers, the elaborate system of cross-referencing plates, numbers, and letters was not so much informational as it was a circular endgame of "non-sense": a pleasant riddle or a parlor game that amused participants. For those who could visit the garden and purchased the folio, the plates and text provided a virtual whirlwind tour of all the amusements on display.

Playfulness is traditionally considered antithetical to the aesthetic goals of the Picturesque, but it provides an essential insight into the dissemination of the style, particularly in the urban milieu. Strollers returned to the garden because the space was intentionally amusing and the pursuit of surprise stimulated both mind and body. The garden stroll offered emotional exhilaration in a newly constructed virtual reality—the Picturesque as gamescape—offering both excitement and entertainment. Considering Monceau in this way suggests that contemporaries recognized that the garden was an intensely social interactive game.

In the early 1770s a number of venues were specifically dedicated to strolling as a form of entertainment. In 1771, before Chartres asked Carmontelle to intervene in his garden, Simon-Gabriel Boutin (1720–94) opened his own *folie*, called Tivoli, to the public.[37] Boutin created a series of interlocking areas within the garden, which were devised to amuse and astonish visitors. Tivoli did not offer the same array of luxury goods or costumed servants as the garden at Monceau, but it became one of the most sought after Picturesque gardens for promenading during the Revolutionary decade. Less than a kilometer from Monceau, and next to Tivoli, two of the most celebrated pyrotechnicians of the era, Claude Ruggieri and Giovanni Battista Torré, offered fireworks on a weekly basis.

Chartres may have turned his attention away from the overt playfulness of Monceau in the 1780s when he hired Thomas Blaikie to redesign parts of the garden. In 1792, after assuming the name Philippe-Égalité, he donated the property to the Republic. The gesture did not save his life but testifies to the fact that he recognized that Monceau was intended as a place for strolling and entertainment that encouraged social mingling.[38] In 1794 the Revolutionary government, specifically the *Comité de Salut Public*, proposed that Monceau could become a school for agriculture. Despite the hoped-for pedagogical value of this project, financial pressure encouraged the municipality of Paris to rent the buildings to a certain Blanchard, who transformed the site into a *jardin-spectacle*. In 1798 Carmontelle's gamescape had become a place to stroll while watching astonishing events for a small fee. These included fireworks in the Temple of Psyche, dancers in Turkish costumes, pantomimes, explorations of the winter garden and its grotto, and balloon ascensions. By 1799 Monceau could not rival its neighbor Tivoli, but more than any other venue it contributed to the metamorphosis of the Picturesque into a *jardin-spectacle*, a precursor to modern amusement parks.

NOTES
Carmontelle and His World
Translated from the French by Andrew Ayers

1. [François] Guizot. *Mémoires pour servir à l'histoire de mon temps par M. Guizot*, vol. 1 (Paris: Michel Lévy Frères, 1870), 6.

2. Unless otherwise specified, background information on Carmontelle comes from Laurence Chatel de Brancion, *Carmontelle au jardin des illusions* (Château de Saint-Rémy-en-l'Eau: Éditions Morelle Hayot, 2003).

3. The grandson of the duc de Luynes having been killed during the war, his heir was his grandnephew, the vidame d'Amiens, married to his granddaughter.

4. *Théâtre de campagne* (Paris, 1775), and see note 2.

5. "Principales époques relatives à Monseigneur le duc de Chartres," K 142, fol. 18, Archives Nationales, Paris.

6. "Principales époques relatives à Monseigneur le duc de Chartres."

7. Charles Maurice, duc de Talleyrand-Périgord, *Mémoires du prince de Talleyrand*, vol. 1 (Paris: Calmann Lévy, 1891), 149.

8. Jean-Aymar Piganiol, seigneur de la Force, *Description historique de la Ville de Paris et de ses environs*, new ed. (Paris: Les Libraires Associés, 1765), 2:329.

9. William Lucas, *A Five Weeks Tour to Paris, Versailles, Marl. &c.* (London, 1765), 21; William Cole, *Journal of My Journey to Paris* (London, 1931), 258.

10. *Mémoires inédits de madame la comtesse de Genlis sur le dix-huitième siècle et la Révolution française, depuis 1756 jusqu'à nos jours* (Paris, 1824), 1:315–16. The comtesse de Genlis was the duc de Chartres's mistress. Her aunt, the marquise de Montesson, was first the duc d'Orléans's mistress and then his wife. The marriage was permitted by the king on the condition that it remain morganatic and the marquise never appear at court.

11. Carmontelle, *Proverbes dramatiques* (Paris: Chez Merlin, 1768), 1:121, 1:123.

12. The French propertied gentry had decided that gentlemen would wear a specific color at their estates. The duc d'Orléans chose red for the *habit de campagne* at Saint-Cloud and green at Villers-Cotterêts. Carmontelle mocked this usage in a comedy, *L'Uniforme de campagne* (1781).

13. Denis Diderot, *Œuvres de Denis Diderot: Salons*, vol. 1 (Paris: Chez J. L. J. Brière, 1821), 51.

14. *Lettres de la marquise du Deffand à Horace Walpole écrites dans les années 1766 à 1780, auxquelles sont jointes des lettres de Madame du Deffand à Voltaire dans les années 1759 à 1775. Publiées d'après les originaux* (Paris, 1812), 1:207n3.

15. Sale, Sotheby's, Paris, September 29–30, 2015, no. 6, fig. 1 (reproducing note, possibly by Richard de Lédans, on the back of Carmontelle's *Les Gentilshommes du duc d'Orléans dans l'habit de Saint-Cloud*).

16. Reproduced in Chatel de Brancion, *Carmontelle*.

17. Denis Diderot, *Correspondence* (Paris: Les Éditions de Minuit, 1955–1970), 3:115.

18. Emilie de Breteuil, marquise du Chatelet (1706–49), translated Newton's *Principia Mathematica*, publicized Leibniz's work, and explored kinetic energy. She and Voltaire had a long affair, and she housed him at her Château de Cirey from 1734 until her death in the aftermath of childbirth in 1749.

19. Quote from Charles Perrault, *La Barbe Bleue* (1697), paraphrased in the song, "Marlborough s'en va-t-en guerre." Beaumarchais would include it in *Le Mariage de Figaro* in 1781.

20. *Correspondance littéraire, philosophique et critique par Grimm, Diderot, Raynal, Meister*, ed. Maurice Tourneux (Paris: Garnier Frères, 1877–88), 4:320.

21. [Mathieu-François Pidansat de Mairobert, ed.], *Mémoires secrets pour servir à l'histoire de la république des lettres en France depuis 1762 jusqu'à nos jours* (London: Chez John Adamson, 1784–1789), 18:221. The success of the opera precipitated a violent quarrel over music theory between Rousseau and France's preeminent composer, Jean-Philippe Rameau (1683–1764). In his entries on music in the *Encyclopédie*, Rousseau had attacked the French musical traditions typified by the music of Rameau.

22. Pidansat de Mairobert, *Mémoires secrets*, 4:165.

23. "Notice sur M. de Carmontelle," Archives de l'Opera, Paris.

24. Marie Paule Angélique d'Albert de Luynes, duchesse de Chaulnes (1744–?) was the daughter of Marie Charles Louis d'Albert de Luynes (1717–1771), duc de Chevreuse (1735), fifth duc de Luynes (1768). In 1758 she married her cousin Louis Joseph d'Albert d'Ailly, vidame d'Amiens, Carmontelle's pupil, who assumed the titles of duc de Picquigny (1762) and duc de Chaulnes (1769). He abandoned her on their wedding night, instead traveling to Egypt from where he brought back the first known drawings of the pyramids of Djoser in Saqqara.

25. Tourneux, *Correspondance littéraire*, 5:282–83.

26. Carmontelle had written several of his dramatic proverbs for Lallemant de Betz.

27. "The king was about to buy it for his library. The author would prefer to see his drawings exiled rather than dispersed. M. de Carmontelle's manner is well known: all his drawings are full-length and in color . . . It would be a witty, amusing acquisition" (Diderot, *Correspondence*, 4:214, 4:220).

28. Tourneux, *Correspondance littéraire*, 5:282.

29. Letter from Leopold Mozart to his wife, April 1, 1764, in Wolfgang Amadeus Mozart, *Correspondance*, vol. 1 (Paris: Flammarion, 1986), 91.

30. Tourneux, *Correspondance littéraire*, 16:352.

31. Ms 9446, Bibliothèque de l'Arsenal, Paris.

32. Carmontelle, *Théâtre de campagne* (Paris: Chez Ruault, 1775), 1:7.

33. *La Promotion*, Carmontelle, *Conversations des gens du monde dans tous les temps de l'année* (Paris: Imprimerie Polytype, 1786), 1:111–12.

34. *Le Sourd*, Carmontelle, *Proverbes dramatiques* (Paris: Chez Merlin, 1768), 1:53.

35. *Le Bal*, Carmontelle, *Proverbes dramatiques* (Paris: Chez Merlin, 1768), 2:158. See Laurence Chatel de Brancion, *Carmontelle's Landscape Transparencies: Cinema of the Enlightenment* (Los Angeles: J. Paul Getty Museum, 2008), 74.

36. An album of illustrated dramatic proverbs is kept at the Musée Condé in Chantilly. Other dramatic proverbs accompanied by Carmontelle's set designs are in private hands. See Chatel de Brancion, *Carmontelle au jardin des illusions*, 150.

37. [François-Auguste Fauveau], baron de Frénilly, *Souvenirs du Baron de Frénilly* (Paris: Plon-Nourrit, 1908), 7–8.

38. Frénilly, *Souvenirs*, 7.

39. Henry Tronchin, *Un Médecin du XVIIIème siècle, Théodore Tronchin, 1709–1781, d'après des documents inédits* (Paris, 1906), 321, 323.

40. Tourneux, *Correspondance littéraire*, 9:263.

41. Frénilly, *Souvenirs*, 6–7.

42. Alden R. Gordon, "L'hôtel Buizette puis Marigny au Roule" in *Rue du Faubourg-Saint-Honoré*, ed. Béatrice de Andia and Dominique Fernandès (Paris: Délégation à l'Action artistique de la ville de Paris, 1994), 392 et seq. The house, today destroyed, stood at 48, rue du Faubourg du Roule, on a plot whose current address is 188, rue du Faubourg Saint Honoré.

43. Pidansat de Mairobert, *Mémoires secrets*, 7:20.

44. Although Carmontelle describes the Naumachia as antique, it seems probable that its columns came from a never-completed Valois funerary chapel that had been commissioned by Catherine de Médicis for Saint-Denis. By the early eighteenth century, the chapel was on the verge of collapse. The Regent, the duc d'Orléans, ordered its dismantling in 1719 and may have come into possession of the columns then; that is the version of events given by Alexandre Lenoir in his *Musée des monuments français* in 1801. See Alexandre Lenoir, *Musée des monuments français*, (Paris: De l'Imprimèrie de Guilleminet, l'An X–1801), 2:31, and A. de Boislille, "La sépulture des Valois," *Mémoires de la Société de l'histoire de Paris et et de l'Ile de France* 4 (1877), 241–92. On the other hand, Michele Beth Bassett asserts that this attribution is false in her thesis "The Funerary Patronage of Catherine de Medici: The Tomb of Henri II, Heart Monuments and the Valois Chapel" (PhD diss., Columbia University, 1999). Her argument, however, does not necessarily stand up to closer scrutiny. She writes (page 182) that "Louis XIV ordered the démolition of the Valois chapel on 24th March 1719," with a footnote indicating "Letters patent, Louis XIV, Paris 24 March 1719" as the source. But Louis XIV had died on September 1, 1715. Louis XV was nine years old at the time, and it was the Regent who was ruling the realm. Two facts seem to support the possibility that it was he who bought the Valois columns. First, a considerable quantity of columns used at Monceau do not appear in the Orléans payment ledgers, meaning that they must have come from the family's reserves. Second, if the Valois columns weren't sold immediately in 1719, it would have been difficult to find a buyer for them shortly thereafter, given the 1720 financial crash associated with the bursting of the South Sea bubble. Given the Regent's vast personal wealth, it would have been easy for him to spend a few thousand livres to acquire the columns. Those used in the Naumachia and the circular temple at Monceau could well be those of the interior chapels, which Bassett herself acknowledges.

45. Carmontelle would have understood the need to keep the water moving, given that the malodorous waters of the Colisée had led to its demise. See Gabriel Wick's contribution in this collection.

46. The King of Poland, Stanislas-Auguste, had sought to acquire it, but Pigalle had preferred to see it remain in France. *Cor-*

respondance inédite du roi Stanislas Auguste Poniatowski et de Madame Geoffrin (1764–1777), ed. Charles de Mouy (Paris: E. Plon et Cie, 1875), 283; Prosper Tarbé, *Vie et œuvre de Jean-Baptiste Pigalle* (Paris: Ve J. Renouard, 1859), 100. Today the sculpture is in the Musée du Louvre.

47. Carmontelle's description of the female African attendant as "bronze" is incorrect; she was in fact made of lead, which was painted black according to Luc-Vincent Thiéry's description from 1787. Carmontelle does not specify the metal or finish of the ewer, but Thiéry described it as "gold." [Luc-Vincent] Thiéry, *Guide des amateurs et des étrangers voyageurs à Paris* (Paris: Chez Hardouin & Gattey, 1787), 1:69. Thanks to the Altman bequest in 1913, the marble bather is now in the collections of New York's Metropolitan Museum. It seems that there were two statues of African women at Monceau, according to an inventory made by Jean-Baptiste-Pierre Lebrun at Monceau after Orléans's condemnation (Inventory of 23–24 fructidor of the Year 2, in the *Procès-verbaux de la Commission des arts*, [Paris, 1917], 2:354–55). The one that was part of the fountain group by Houdon "had been mutilated by scoundrels"; another, of unknown authorship, was "in very good condition" inside the Great Pyramid. The lead statues as well as all the lead objects may have been melted down to manufacture munitions for France's Revolutionary wars. A bronze cast of a terracotta bust that may have been the model for the African attendant was created by Houdon in 1794 to celebrate the abolition of slavery. It can be found in the Musée Nissim de Camondo in Paris, and there is a smaller bronze version in the Musée des Arts Décoratifs in the same city (Louis Réau, *Houdon: sa vie et son œuvre* [Paris: F. de Nobele, 196_], 236 et seq.) A plaster version of the head is in the collection of the Musée Municipal de Soissons (Tanya Paul, "Houdon's Bather in a Drawing by Piere Antoine Mongin," *Metropolitan Museum Journal*, vol. 48 [2013]: 161–67; Anne L. Poulet, *Jean-Antoine Houdon: Sculptor of the Enlightenment*, exh. cat. [Washington, DC: National Gallery of Art in association with the University of Chicago Press, 2003]).

48. Alden R. Gordon, *The Houses and Collections of the Marquis de Marigny* (Los Angeles: Provenance Index of the Getty Research Institute, 2003), 45–46, and Gordon, "L'hôtel Luizette puis Marigny," 392 et seq. The catalogue of the 2016 Bouchardon exhibition at the Louvre and Getty Museums (the faun is no. 14) erroneously states that the king gave the work to Marigny in 1753 for Monceau; Monceau did not exist at the time and never belonged to Marigny. Bouchardon's sculpture was confiscated with all the other Orléans possessions in 1793. Pajou provided confirmation of its provenance to the Comité d'Instruction publique (Comité d'Instruction publique, D XXXVIII, Archives Nationales, Paris). Today Bouchardon's faun is in the collections of the Louvre, while the original antique statue was bought by the king of Bavaria sometime after 1810 and is now in the Glyptothek in Munich.

49. Quoted in Morris R. Brownell, *Alexander Pope and the Art of Georgian England* (Oxford: Clarendon Press, 1978), 245.

50. Letter to his daughter Sukey, October 23, 1762, cited in Kenneth Woodbridge, *Landscape and Antiquity: Aspects of English Culture at Stourhead 1718–1838* (Oxford: Clarendon Press 1970), 31.

51. Denis Diderot, *Œuvres completes de Diderot revues sur les éditions originales*, ed. by J[ules] Assézat (Paris: Garnier Frères 1875), 5:87; 6:65.

52. DQ 346, Archives de Paris. Lenoir founded a museum to preserve ecclesiastical, aristocratic, and royal sculpture and monuments that were threatened by iconoclasm during the Revolution. Ultimately the museum was housed in a decommissioned convent.

53. Carmontelle, "Mémoire sur les tableaux transparents du citoyen Carmontelle, l'an 3e de la Liberté" (1794). Autographes 008.1, Bibliothèque de l'Institut national d'histoire d'art, Paris.

54. Voltaire, *Mercure de France*, June 1745, part 1, 13.

55. On the question of Freemasonry at Monceau, see David L. Hays, "Carmontelle's Design for the Jardin de Monceau: A Freemasonic Garden in Late-Eighteenth-Century France," *Eighteenth-Century Studies* 32, no. 4 (Summer 1999): 447–62; Jay Macpherson, "Masonic Landscape Design, or Down the Garden Path" in *Ars Quatuor Coronatorum* 110 (1997); Chatel de Brancion, *Carmontelle au jardin des illusions*, 134–136.

56. Louis-Philippe de Bourbon, whose title at birth was duc de Valois, would become King of the French in 1830.

57. *Mémoires de la baronne d'Oberkirch sur la cour de Louis XVI et la société française avant 1789*, ed. Suzanne Burkard (Paris: Mercure de France, 1989), 470 et seq.

58. Emmanuel de Croÿ-Solre, *Journal inédit du duc de Croÿ (1718–1784)* (Paris: Flammarion, 1907), 3:211.

59. Émile Dacier, "Une peinture inconnue de Saint Aubin: La Naumachie et les jardins de Monceau," *Revue d'Art ancien et moderne* 25 (1909): 207, and "Le jardin de Monceau avant la Révolution" *Revue de la Société d'iconographie parisienne*, troisième année (1910): 56.

60. Gabriel de Broglie, *Madame de Genlis* (Paris: Perrin, 2001), 63.

61. Frénilly, *Souvenirs*, 7.

62. Tourneux, *Correspondance littéraire*, 11:371–72.

63. Tourneux, *Correspondance littéraire*, 11:372–74.

64. Tourneux, *Correspondance littéraire*, 11:375–76.

65. On the corrupting influence of marquis de Voyer on the duc de Chartre, see Talleyrand-Périgord, *Mémoires*, vol. 1, 152ff.

66. Charles-Joseph Lamoral, prince de Ligne, *Mémoires, lettres et pensées*, ed. Alexis Payne (Paris: Éditions François Bourin, 1989), 449.

67. At the beginning of 1775, Louis XVI had ordered the felling of all old growth and declining trees at Versailles. Hubert Robert memorialized the "unique and frightening spectacle" in a painting now at the Musées de Versailles et du Trianon; see Pidansat de Mairobert, *Mémoires secrets*, 11:31.

68. After a career on the battlefield, François-Henri-René, duc d'Harcourt, became ambassador to Great Britain, where he discovered the parks of English country states. Carmontelle may have met this great lover of theater in the houses of actresses. Fragonard, who frequented many of the same houses, painted a fine portrait of the duke, currently in Musée du Louvre.

69. *Essay on Modern Gardening by Horace Walpole, With a Faithful Translation into French by the Duke of Nivernois: A Reprint in Type Facsimile of the Edition Printed by Mr. Walpole at Strawberry Hill MDCCLXXXV*, ed. Alice Morse Earle (Canton, PA: Lewis Buddy III–The Kirgate Press, 1904), 4–5.

70. Some of the drawings had already been engraved, as noted in the *Correspondance littéraire* of November 1776. Tourneux, *Correspondance littéraire*, 11:372.

71. Thomas Blaikie, *Diary of a Scotch Gardener at the French Court at the End of the Eighteenth Century*, ed. Francis Birrell (Cambridge: Cambridge University Press, 2012), 179. [Ed. nt. Blaikie's spelling has been silently corrected.]

72. École Nationale des Beaux Arts, Paris, reproduced in Chatel de Brancion, *Carmontelle au jardin des illusions*, 168.

73. *Le Jardin anglais* was first published by Clarence D. Brenner in "Le développement du proverbe dramatique en France et sa vogue au XVIIIè siècle, avec un proverbe inédit de Carmontelle," *University of California Publications in Modern Philology* 20, no. 1 (1937), 1–52.

74. Frénilly, *Souvenirs*, 8.

75. Sale, Sotheby's, Paris, September 29–30, 2015, nos. 43 and 44.

76. On emakimono, see Miyeko Murase, *Emaki Narrative Scrolls from Japan* (New York: The Asia Society, 1989), 15–16.

77. "Pour que les objets présents sur cette bande papier passent successivement, elle est montée sur deux rouleaux de bois enfermés dans une boite noire qu'on a placés à ses extrémités" (so that the objects present on this strip of paper come into view successively, it is mounted on two wooden rollers enclosed in black boxes that are placed at its extremities), "Mémoire sur les tableaux transparents du citoyen Carmontelle, l'an 3è de la Liberté."

78 Frénilly, *Souvenirs*, 8.

79. Frénilly, *Souvenirs*, 8.

80. Carmontelle, *Conversations des gens du monde dans tous les temps de l'année* (Paris: A l'imp. Polytype, 1786), unpaged preface.

81. Germaine de Staël, *Correspondance générale*, ed. Béatrice W. Jasinski, vol. 1 (Paris: J.-J. Pauvert, 1962), 112 (letter of August 9, 1786).

82. Thomas Crow, *Painters and Public Life in Eighteenth-Century Paris* (New Haven: Yale University Press, 1985), 18.

83. Diderot, *Salon de 1765* (Paris: Hermann, 1984), 87–88.

84. *Coup de Patte sur le Sallon de 1779, dialogue; précédé et suivi de réflexions sur la peinture* (Athens and Paris, 1779), 1.

85. *Coup de Patte sur le Sallon de 1779*, 7.

86. *Le Triumvirat des Arts, ou Dialogue entre un peintre, un musicien et un poète sur les tableaux exposés au Louvre année 1783 pour servir de continuation au Coup de Patte et à la Patte de Velours* (Aux Antipodes, 1783), 3–4.

87. *Le Frondeur, ou Dialogues sur le Sallon, par l'Auteur du Coup-de-Patte et du Triumvirat* (1785), 3.

88. *Le Frondeur*, 20. Carmontelle was referring to Vien's *Retour de Priam avec le corps d'Hector* (Return of Priam with the body of Hector), exhibited at the Salon of 1785 (no. 1 of the catalogue). It depicted the Trojan king greeted by members of his family, including Hecuba, Andromache, Cassandra, and the child Astianax. Paris and Helen, "fearing reproaches," stood a little apart from the others. *Explication des peintures, sculptures et gravures, de messieurs de l'Académie royale* (Paris: De l'Imprimerie de la Veuve Hérissant, 1785), 3–4.

89. *Vérités agréables ou le Salon vu en beau, par l'auteur du Coup-de-patte* (Paris, 1789), 23, 3.

90. *Vérités agréables ou le Salon*, 23.

91. Six Salon critiques published anonymously between 1779 and 1789 were meant to be understood as being by the same author. They were: *Coup de Patte sur le Sallon de 1779, dialogue; précédé et suivi de réflexions sur la peinture* (Athens and Paris: Chez Cailleau, 1779); *La Patte de Velours, pour servir de suite à la seconde édition du Coup de Patte; ouvrage consernant le Sallon de peinture, année 1781* (London and Paris: Chez Cailleau, [1781]); *Le Triumvirat des Arts, ou Dialogue entre un peintre, un musicien et un poète, sur les tableaux exposés au Louvre, année 1783. Pour servir de continuation au Coup de Patte et à la Patte de Velours* (Aux Antipodes, [1783]); *Le Frondeur, ou Dialogues sur le Sallon, par l'auteur du Coup-de-Patte et du Triumviart* (1785); *Encore un Coup de Patte, pour le dernier, ou Dialogue sur le Salon de 1787* (1787); *Vérités agréables ou le Salon vu en beau, par l'auteur du Coup-de-patte* [Paris, 1789]. The *Mémoires secrets* of October 6, 1781, as well as *Le Pourquoi ou l'ami des artistes* (Geneva, 1781), 10–11, indicated that Carmontelle was their author. However, this assignment was called into question in an obituary of Louis-Henri Lefébure (1754–1839), a former subprefect of Verdun. The obituary was written in 1839 by a Monsieur Voiart, "despite his old age," for the *Mémoires de la Société royale des sciences, lettres et arts de Nancy* (1840), 233–34. The elderly Voiart claimed Lefébure himself as his informant: "The information I give you comes mostly from the mouth of the man of whom I write . . . So be indulgent, gentlemen, because memory alone is the source from which I draw." Concerning the Salons, Voiart wrote that a series of articles appeared "that were published under the title *Coup de Patte sur le Salon de 1788*. "The discriminating criticism and excellent taste of this small work proved that its author was no stranger to the art of painting." Voiart mentions only this booklet, and the worth of his testimony is called into question by the fact that there was no Salon in 1788 and that the obituary contains many implausible statements. However, Michaud in his *Biographie universelle ancienne et moderne: histoire par ordre alphabétique de la vie publique et privée de tous les hommes* ([Paris, 1843], 23:577), followed by Firmin-Didot, in *Nouvelle biographie générale depuis les temps les plus reculés jusqu'à nos jours* ([Paris, 1854–66], 30:304), attributed the six pamphlets to Lefébure. Their assertions were repeated in turn by Helena Zmijewska in her *La Critique des salons en France du temps de Diderot* and Richard Wrigley in "Ways of Seeing at the Salon," *Art History* (September 1982), and "Censorship and Anonymity in Eighteenth-Century France Art Criticism," *Oxford Journal of Arts* (February 1983), and supported by Wrigley in his *The Origins of French Art Criticism. From the Ancient Régime to the Restoration* (Oxford: Clarendon Press, 1993), app. 4. However, other contemporary scholars have favored the attribution of the writings to Carmontelle, among them Régis Michel in *David et Rome* (Rome: French Academy of Rome, 1981) and "Diderot et la modernité" in *Diderot et l'art de Boucher à David* (Paris: Éditions de la Réunion des musées nationaux, 1984); Annie Beck in *Lumières et modernités: de Malebranche à Baudelaire* (Paris, 1994); I myself in *Carmontelle au jardin des illusions*; and Thomas Crow in *Painters and Public Life in Eighteenth Century Paris* (New Haven: Yale University Press, 1985) (except perhaps for the pamphlet relating to the Salon of 1787). (Crow has since equivocated with reference to the pamphlet of 1789, writing that "the attribution of this pamphlet remains undertain" [*Emulation: Making Artists for Revolutionary France* (New Haven: Yale University Press, 1995), 317n70]). Pierre Vaillandet, in "Rapports de Louis-Henri Lefébure, Commissaire du pouvoir exécutif en Vaucluse (1793-an II)," in *Annales d'Avignon et du comtat Venaissin* (Avignon, 1932), 69–136, revealed that Lefébure spent his youth with his family in Provence and worked later in Paris as a "Maître de musique," never being involved professionally with the fine arts. Furthermore, Lefébure's 1821 "complete list" of his writings does not include the titles in question (F/1B1/166/20, Archives Nationales, Paris). This provides strong support for the attribution to Carmontelle.

92. "Mémoire sur les travaux transparents du citoyen Carmontelle, l'an 3ᵉ de la Liberté," and a transparency titled *Les Quatre Saisons* (The Four Seasons), now in the collections of the Musée du domaine départemental de Sceaux. Both are reproduced in Chatel de Brancion, *Carmontelle's Landscape Transparences*, 129–30 and 92–107.

93. "Projet par Carmontelle de revêtement pour les murs de la terrasse des Tuileries, 22 frimaire an III," F/17/1344/35, Archives Nationales, Paris.

94. The manuscripts were bound together and were long thought lost. They came on the market in 2018 and acquired by the Institut national d'histoire de l'art, Paris, in November of that year.

95. Jeaurat's *Traité de perspective à l'usage des artistes de l'ingénieur-géographe* and Lambert's *La Perspective affranchie de l'Embarras du Plan géométral*.

96. Carmontelle, "La Perspective démontrée à l'usage des jeunes gens qui savent la géométrie et le dessin" (1794–95). Ms. 845, Bibliothèque de l'Institut national d'histoire de l'art, Paris.

97. Carmontelle, "La Perspective démontrée."

98. Carmontelle, "La Perspective démontrée."

99. *Procès-verbaux de la Commission des arts*, 1:196.

100. "Décret de la Convention du 14 floréal an II," in *Procès-verbaux de la Commission des arts*, 1:506.

101. Marc Furcy-Raynaud, "Les tableaux et objets d'art saisis chez les émigrés et les condamnés et envoyés au Muséum central," *Archives de l'art français*, nouvelle période, 6 (1912): 75.

102. See the Mellon Collection transparency reproduced in Chatel de Brancion, *Carmontelle's Landscape Transparencies*, 116–24.

103. *Proverbes et comédies posthumes de Carmontel* [*sic*], ed. [Stéphanie Félicité], comtesse de Genlis, vol. 1 (Paris: Chez Ladvocat, 1825), vi.

104. During the construction of the tax wall around Paris in the 1780s, Ledoux built fifty-five neoclassical toll houses at each of its entry points. The one at Monceau was in fact a surveillance post, the duc d'Orléans having seen to it that no wall would be built along his garden and that the original ha-has would remain in place: people had to go around the park to enter the town.

105. The hôtels de Camondo, Cernuschi, and André have all become museums displaying their owners' collections.

106. Albert Wolff, *La Capitale de l'art*, 2nd ed. (Paris: Victor-Havard Éditeur, 1886), 285.

107. *Quarelling*, 1874 (private collection); *The Convalescent*, 1876 (Sheffield Museum, UK); *Holiday*, 1876 (Tate Britain, London); *The Hammock*, 1879 (private collection).

108. Monet produced six views of the Parc Monceau (four in oil on canvas and two in pastel). Two are now in New York's Metropolitan Museum of Art.

109. The construction of the statue began in the mid-1870s at the foundry owned originally by Honoré Monduit, who was responsible for the statue's right hand, torch, and flame. After he retired, the rest of the statue was completed by his younger partners, Emile Gaget and J. G. Gauthier.

History by Design: The Aesthetics of Transformation in Carmontelle's Jardin de Monceau

1. "Inventaire des Titres qui constituent Les Jardins de Mousseaux appartenans à S. A. S. Monseigneur Le Duc d'Orléans," 300 AP I (184), fol. 7r, Archives Nationales, Paris (hereafter cited as AN). Little is known of Colignon's career. Among the works in Paris attributed to him are a house on the rue d'Argenteuil, just west of Palais-Royal, 1762; the hôtel de La Vaupalière, 1768, at the southwest corner of the rue du Faubourg Saint-Honoré and the avenue Matignon; and the hôtel de Traverse, rue Nôtre Dame-des-Champs, 1776. See Michel Gallet, *Paris Domestic Architecture of the 18th Century*, trans. James C. Palmes (London: Barrie & Jenkins, 1972), 151.

2. On the management of the *couronne*, see Pierre Gaxotte, *Paris au XVIIIe siècle* (Paris: Arthaud, 1968), 60–61. A declaration of July 1724 prohibited any new construction there (Gallet, *Paris Domestic Architecture*, 1). Also, Louis XIV's plan of 1670 for an open but still defensible Paris required that the terrain beyond the boulevards remain barren or in cultivation. Exterior fortifications ringing the city protected the urban core, but the area between the boulevards and those fortifications was within range of enemy cannons.

3. See Durand Echeverria, *The Maupeou Revolution: A Study in the History of Libertarianism, France 1770–1774* (Baton Rouge and London: Louisiana State University Press, 1985), 10; Gallet, *Paris Domestic Architecture*, 9.

4. "Déclaration par Jean Flagère," AN, 300 AP I (184). See David L. Hays, "'This Is Not a *Jardin anglais*': Carmontelle, the Jardin de Monceau, and Irregular Garden Design in Late-Eighteenth-Century France," in *Villas and Gardens in Early Modern Italy and France*, ed. Mirka Beneš and Dianne Harris (New York: Cambridge University Press, 2001), 410–11n36.

5. "Vente à vie par Marie Louis Colignon," AN, 300 AP I (184). See also "Vente à vie de 3 arpents 1/2 par Marie Louis Colignon, architecte et entrepreneur des bâtiments du roi, rue du Four, à Louis-Philippe-Joseph d'Orléans, duc de Chartres, d'un terrain près de la barrière de Monceau," AN, MC/RS/1078 (formerly MC/ET/LIII/459). The design agreed upon was described in the contract and represented by an attached plan of the whole property plus plans and elevations of the pavilion. According to the contract (300 AP I (184), page 3, recto; MC/RS/1078, page 19, recto), the garden was to be "distribué en avenues d'ormes formant Étoile, aboutissant sur chacune des huit faces du dit bâtiment, les massifs seront variés par différentes Salles et bosquets de verdure" (distributed in avenues of elms forming a star, terminating at each of the eight faces of the building, the clumps [of vegetation] will be varied by different rooms and groves of greenery). See also Hays, "'This Is Not a *Jardin anglais*,'" 299. In a *vente à vie* transaction, the rights to the property were sold for a single

price, although the payments could be made over time. On the death of the purchaser or a default on payments, the property would return in full to the estate of the original owner.

6. The pavilion was evidently finished by or before July 24, 1771, when the duc de Chartres hosted a dinner there for some of his friends ("Annectdottes Galantes du 2 aoust 1771," in "Journal de Police 1766–1777," Manuscrits FR 11360, Bibliothèque Nationale, Paris). In March 1773 the duc de Chartres paid Colignon 3,000 livres along with a bonus of 600 livres for speed of execution. See "Mémoire sur Monceau concernant le Sr. Colignon," AN, 300 AP I (184), fol. 3r.

7. The circumstances surrounding this acquisition were highly complex. On May 29, 1771, Emery and Dessaudé agreed to rent a lot of around 3.5 acres (1.4 hectares) to Philippe Duchaine, *dit* St. Denis, and his wife, Appoline Hamoche (parcels B1, B2, and B3 in fig. 25). The terms involved a *bail emphytéotique* of eighteen years to start on June 1, 1771, with an initial payment—a *pot de vin*—of 1,200 livres and an annual rent of 800 livres. Four weeks later, on June 28, 1771, St. Denis and Hamoche signed a *sous-bail emphytéotique* in which they agreed to sublet parcel B1 for the duration of their own tenancy and at an annual rate of 500 livres to Jean-Baptiste Maublanc, a *valet de chambre* on the duc de Chartres's household staff, and his wife, Catherine Elizabeth Daler, who were secretly acting as the duke's proxies. Matters became even more complicated when, on December 27, 1773, Emery and Dessaudé signed a contract in which they reserved future tenancy of the entire area (B1, B2, and B3) to the duc de Chartres; specifically, they agreed to a *bail à vie* to take effect upon the expiration of the *bail emphytéotique* to St. Denis and Hamoche (i.e., after June 1, 1789). While the annual rent for the area was to remain consistent with that paid by St. Denis and Hamoche—800 livres divided into semiannual payments—Emery and Dessaudé also demanded a *pot de vin* of 8,500 livres from the duc de Chartres, payable immediately. With these agreements and payments in place, Emery and Dessaudé created a great deal of confusion when they then decided to sell full title to parcels B2 and B3 to their tenants St. Denis and Hamoche. That arrangement—transacted on July 16, 1774—undermined the promise of future tenancy by the duc de Chartres agreed upon seven months before. To protect his interest, the duke was present at the signing of the sale to St. Denis and Hamoche. Emery and Dessaudé continued to own parcel B1, already developed by the duc de Chartres, and St. Denis and Hamoche ultimately retained parcel B3 for themselves, but they reduced their costs by immediately ceding parcel B2 to the duc de Chartres with a *bail à vie* and by demanding a *pot de vin* of 4,250 livres to be paid by Emery and Dessaudé from the money paid to them earlier by the duc de Chartres to secure the whole property. As part of the final transaction, St. Denis and Hamoche agreed to be responsible for tearing down the large stable and riding ring which occupied parcel B2, leaving the duc de Chartres to develop the property as he pleased with the understanding that, upon expiration of the contract, all immovable property situated there would become property of St. Denis and Hamoche or their heirs. See "16 Juillet 1774," AN, 300 AP (184).

8. On a baptismal certificate in 1744, Carmontelle identified himself as an *ingénieur* (A. Augustin-Thierry, *Trois amuseurs d'autrefois: Paradis de Moncrif, Carmontelle, Charles Collé* [Paris Librairie Plon-Nourrit et Cie., 1924], 80.) In 1757, Emmanuel-Louis-Auguste de Pons Saint-Maurice, commander of the Orléans dragoons and governor of the young duc de Chartres, hired Carmontelle as an aide-de-camp and *officier ingénieur* for a military campaign to Westphalia (Amédée Britsch, *La Maison d'Orléans à la fin de l'ancien régime. La Jeunesse de Philippe-Égalité* [Paris: Payot, 1926], 21; Thierry, 81, 83.)

9. See, for example, Auguste-François-Fauveau, baron de Frénilly, *Souvenirs du Baron de Frénilly, Pair de France (1768–1828)*, ed. Arthur Chuquet (Paris: Librarie Plan, 1909), 7.

10. Parcels C and D were obtained from Emery and Dessaudé on December 17, 1772, and March 13, 1773, respectively. Maublanc and Daler then transferred those parcels to the duc de Chartres on February 29, 1776. Parcel C covered about 0.44 acre (0.18 hectare) and cost 2,850 livres, whereas parcel D covered less than a quarter of that area (0.1 acre; 0.04 hectare) and cost 1,500 livres. The grossly ill-proportioned plan drafted to illustrate the transaction of parcel D—elucidated by the known locations of cisterns and by written descriptions of the contiguity of the property with neighboring plots—shows that it was until then the site of a cabaret. The number of extra-urban cabarets and *auberges* set up along travel routes to Paris was considerable throughout the eighteenth century. Historian Bernard Rouleau recorded figures for the Left Bank village of Vaugirard; in 1717, in the village of ninety-five households, there were twenty-seven cabarets (Bernard Rouleau, *Villages et faubourgs de l'ancien Paris* [Paris: Seuil, 1985], 74). Before being purchased by Maublanc and Daler, Parcel D had been rented to a M. Tentard, possibly the *cabaretier* himself; see "Mémoire sur Monceau concernant le Sr. Colignon," AN, 300 AP I (184), 3v. The alignments of parcel E (see fig. 25) also show, importantly, that this cabaret was not the same establishment later found thirty meters to the north and marked on Carmontelle's plan of 1779.

11. In the vicinity of Paris, property developers were required to submit plans of proposed work to the head of the local royal

hunting administration—in this case, the Capitainerie de la Varenne des Tuileries—before enclosing any terrain. That formality was meant to guarantee that new construction would not interfere with the progress of princely hunts. If nothing else, the process set standards for the height and security of property enclosures meant to keep out game. See [Anne-Marie-Édouard-Simon-Stylite-] Y[von] Cazenave de la Roche, *La Venerie royale et le régime des capitaineries au XVIIIe siècle* (Nimes: J. Courrouy, 1926), especially 103–105.

12. [Mathieu-François Pidansat de Mairobert, ed.], *Mémoires secrets pour server à l'histoire de la république des lettres en France, depuis 1762 jusqu'à nos jours* (London: Gregg International Publishers, 1970), 7:18 (June 28, 1773).

13. Louis-Sebastien Mercier referred to the new style of garden as "cette caricature parisienne, qui veut représenter dans quelques arpens, des beautés larges et vraiment originales" (this Parisian caricature, which seeks to represent, in several *arpents*, beauties that are extensive and truly original) (Louis-Sebastien Mercier, *Tableau de Paris, nouvelle edition corrigée et augmentée*, vol. 9 [Amsterdam, 1782], 356.) Thomas Blaikie complained that many of his French patrons failed to appreciate the qualities of landscape gardens. Asked to conceive a garden for the comte d'Artois's newly acquired château at Maisons, Blaikie lamented that he was offered only four or five acres on which to carry out his work. "I told them that was not what was meant by English gardens," he recounted, "that the whole ground round the house ought to correspond else they never could think of having anything beautiful but this they had no ideas of" (Thomas Blaikie, *Diary of a Scotch Gardener at the French Court at the End of the Eighteenth Century*, ed. Francis Birrell [New York: E. P. Dutton & Company, 1932], 132 [September 20, 1777]). [Ed. nt. Blaikie's spelling has been silently corrected.]

14. The architect Jean-François-Thérèse Chalgrin designed a new church building for Saint-Philippe du Roule around 1768. Construction took place between 1773 and 1784. See "Registre pour les délibérations de l'œuvre et fabrique de la paroisse St. Philippe du Roule," AN, HS 3809, and Allan Braham, *The Architecture of the French Enlightenment* (Berkeley and Los Angeles: University of California Press, 1980), 132. The contract with the duc de Chartres (then duc d'Orléans) for parcel E was converted to *bail à vie* on March 22, 1790. See "Inventaire des Titres," article 23, fol. 9v.

15. In general, these were rectangular plots covering at least one *arpent de Paris* (1.26 acres; 0.51 hectares) and planted with scrappy, fast-growing trees and shrubs. On the function and management of the *remises*, see Cazenave de la Roche, *La Venerie*, 127; Jacques Hillairet, *Connaissance du vieux Paris*, vol. 3 (Paris: Éditions Gonthier, 1954), 117; and Yves Gaultier, *La Capitainerie royale des chasses de la Varenne du Louvre* (Paris: Grande Vénerie de France, 1959), 55–58.

16. Full title to parcel G was purchased by Colignon from a certain *sieur* Minoret in an act certified by the notary Le Pot d'Auteuil on June 10, 1774. That same day Colignon sold the lot to the duc de Chartres, in a transaction certified by the same notary, in exchange for 3,600 livres, a "somme excédante de beaucoup, son acquisition" (sum greatly exceeding [the cost of] its acquisition." See "17 Xbre 1772. Vente à Jean Bte Maublanc," AN, 300 AP I (184).

17. On May 28, 1789, the comte de Latouche-Tréville, chancellor of the duke's *conseil*, authorized payment to Colignon of 180,000 livres—a huge sum. Three days later, Colignon signed a statement in which he renounced the fourth article of the original contract and transferred proprietorship of parcel A to his onetime patron. See "Mémoire sur Monceaux Concernant Le s.r Colignon" (completed May 28, 1789) and "Entre Marie Louis Colignon . . . et Pierre Boileau de Berceloup" (May 31, 1789), AN, 300 AP I (184).

18. See note 7, above.

19. On the symbolic meaning of that arrangement, see David L. Hays, "Carmontelle's Design for the Jardin de Monceau: A Freemasonic Garden in Late-Eighteenth-Century France," *Eighteenth-Century Studies* 34, no. 4 (Summer 1999), 447–62.

20. In the *Correspondance littéraire* entry for November 1776, which presented Carmontelle's critique of Jean-Marie Morel's *Théorie des jardins*, both the editor and Carmontelle identified that as a special problem at Monceau. *Correspondance littéraire, philosophique et critique par Grimm, Diderot, Raynal, Meister*, ed. Maurice Tourneux (Paris: Garnier Frères, 1877–88), 11:371–78.

21. Cf. Thomas Whately, *Observations on Modern Gardening* (London: Printed for T. Payne, 1770), 6: "A plain is not, however, in itself interesting; and the least deviation from the uniformity of its surface, changes its nature; as long as the flat [*sic*] remains, it depends on the objects around for all its variety, and all its beauty."

22. No such light line appears on Georges-Louis Le Rouge's later plan of 1783. Georges-Louis Le Rouge, *Jardins anglo-chinois, ou Détails de nouveaux jardins à la mode*, 10ᵉ cahier (Paris: Chez le Rouge, [1783]), pl. 2.

23. Carmontelle wrote that the Bell Tower was "fait exprès pour server de point de vue audessus du bois" (expressly designed to serve as a viewpoint over the wood) (*Garden at Monceau*, 20).

24. Whately framed that difference as a liberation, gardening "being released now from the restraints of regularity" (Whately, *Observations*, 1).

25. L. L. G. D. M., *Lettre sur les jardins anglois* (Paris: Moutard, 1775), 8–9.

26. *Lettre sur les jardins anglois*, 9: "des allées alignées, lesquelles perdant par une gradation nuancée l'appareil de la décoration, se terminent à vos bosquets agrestes."

27. *Lettre sur les jardins anglois*, 10: "plus agréable qu'un purement Anglois"; "infiniment moins destructeur & moins dispendieux."

28. See David L. Hays, "Francesco Bettini and the Pedagogy of Garden Design in Late-Eighteenth-Century France," in *Tradition and Innovation in French Garden Art: Chapters of a New History*, ed. John D. Hunt and Michel Conan (Philadelphia: University of Pennsylvania Press, 2002), 99–101.

29. In his introduction to the French edition of Whately's *Observations on Modern Gardening*, translator François-de-Paule Latapie claimed that the practice of natural design, as exemplified by English example, consisted of imitating or manipulating "les seules beautés de la nature" (the singular beauties of nature) (Thomas Whately, *L'Art de former les jardins modernes, ou, L'Art des jardins anglais*, trans. François-de-Paule Latapie [Paris: Charles-Antoine Jombert, 1771], i–ii).

30. Blaikie began working at Monceau in 1781, primarily developing the greenhouse collections. After a visit to the garden in 1777, he noted in his diary that "there hothouses is not well contrived," and he was sharply critical of Carmontelle's overall design. See Blaikie, *Diary*, 134 (October 3 and 5, 1777), 143 (January 1778), and 178–79 (March 1783).) In January of 1781, Blaikie sent his assistant, Archibald, to England on a mission to buy greenhouse plants for Monceau. He returned in March "with a fine collection." Included were many seeds from Sir Joseph Banks "of those brought home by the Discovery Ships which went out with Captain Cook the 10th Feby 1776 and likewise some Seeds from Mr Morison with a great many other evergreen tree seeds; made a complete Nursery; the only things wanted was a Hothouse for several sorts of seeds from the friendly Islands and Ottahiette [Tahiti]." See January 1781, 168–69, and also December 16, 1779, 160. In 1783 Blaikie assumed the responsibilities of chief plantsman, previously held by "Mr. Scott," and designer, displacing Carmontelle. [Ed. nt. Blaikie's spelling has been silently corrected.]

Fashion Follies at the folie de Chartres

1. Daniel Roche, *The Culture of Clothing: Dress and Fashion in the Ancien Régime*, trans. Jean Birrell (Cambridge: Cambridge University Press, 1994), 159–62.

2. Jean Duma, *Les Bourbon Penthièvre (1678–1793): Une nébuleuse aristocratique au XVIIIe siècle* (Paris: Publications de la Sorbonne, 1995), 69n2.

3. Alexandre Dumas, *Histoire de la vie politique et privée de Louis-Philippe* (Paris: Dufour & Mulat, 1852), 1:4.

4. For a good overview of the "callousness, recklessness, and profligacy" of the duc de Chartres, see Gaston Maugras, *The Duc de Lauzun and the Court of Louis XV* (London: Osgood, McIlvaine, & Co., 1895), 73.

5. Cited in André Castelot, *Philippe-Égalité: Le régicide* (Paris: J. Picollec, 1991), 29.

6. The pressure was particularly great in that a goodly portion of the Orléans's wealth was slated to revert to the state should the line be extinguished for want of male heirs; see Dumas, *Histoire de la vie politique*, 6.

7. On this perception, see Dumas, 4–5; and David L. Hays, "Carmontelle's Design for the Jardin de Monceau: A Freemasonic Garden in Late-Eighteenth-Century France," in *Eighteenth-Century Studies* 32, no. 4 (Summer 1999): 447–62.

8. On Lauzun's disinclination to be faithful to his young bride, see Maugras, *Duc de Lauzun*, 102–5.

9. On Louis XVI's aversion to consummating his marriage, see Caroline Weber, *Queen of Fashion: What Marie Antoinette Wore to the Revolution* (New York: Picador, 2007), 55–56; and Stefan Zweig, *Marie Antoinette: The Portrait of an Average Woman*, trans. Eden & Cedar Paul (New York: Grove Press, 2002), 20–26.

10. Adèle d'Osmond, comtesse de Boigne, *Mémoires de la comtesse de Boigne, née d'Osmond: Récits d'une tante*, ed. Jean-Claude Berchet (1907; repr., Paris: Mercure de France, 1999), 1:43.

11. Émile Langlade, *Rose Bertin: The Creator of Fashion at the Court of Marie-Antoinette*, trans. Angelo S. Rappoport (New York: Charles Scribner's Sons, 1913), 15–16.

12. Weber, *Queen of Fashion*, 103; and "Modes, marchands, marchandes de," in Denis Diderot and Jean Le Rond d'Alembert,

eds., *Encyclopédie, ou dictionnaire raisonné des sciences, des arts et des métiers par une société de gens de lettres* (Lausanne: Chez les Sociétés Typographiques, 1779), 22:18–19.

13. Clare Haru Crowston, "The Queen and Her 'Minister of Fashion'," in *Gender and History*, vol. 14, no. 1 (April 2002): 92–116, 96. See also Weber, *Queen of Fashion*, 102–4; and Madeleine Delpierre, "Rose Bertin, les marchandes des modes et la Révolution," in Catherine Join-Diéterle and Madeleine Delpierre, eds., *Modes et révolutions: 1780–1804*, exh. cat. (Paris: Editions Paris-Musées, 1989), 27–34, 23–24.

14. Being a *marchande de modes* and not a seamstress, Bertin was technically not allowed to make dresses. But in practice she dramatically reconfigured her clients' existing gowns, transforming skirts and sleeves through the ingenious addition of ribbons, flounces, and the like. By the late 1770s she was secure enough in her status as the queen's "minister of fashion" that she openly flouted this trade restriction and began to make dresses instead of merely trimming them.

15. Langlade, *Rose Bertin*, 27; Weber, *Queen of Fashion*, 317n59. On the popularity of the *quesaco*, see also [Jean-Baptiste-Honoré-Raymond] Capefigue, *Louis XV et la société du 18ᵉ siècle* (Brussels: Hauman, 1842), 5:66.

16. [Mathieu-François Pidansat de Mairobert, ed.], *Mémoires secrets de Bachaumont, 1762–1787*, ed. M. J. Ravenel, new ed., vol. 4 (Paris: Brissot-Thivars, 1830), 334–35 (entry of April 26, 1774). Bachaumont specifies that the duchesse de Chartres's new headdress sounded the death-knell for the *quesaco*. On this particular pouf, see also Langlade, *Rose Bertin*, 29; Edmond and Jules de Goncourt, *The Woman of the Eighteenth Century*, trans. Jacques Le Clerq & Ralph Roeder (New York: Routledge, 2013), 235; and Augustin Cabanès and Lucien Nass, *La Névrose révolutionnaire* (Paris: Société Française d'Imprimerie et de Libraire, 1966), 355.

17. Catherine Charlotte, Lady Jackson, *The French Court and Society* (London: Richard Bentley & Son, 1881), 1:97–98.

18. Pidansat de Mairobert, *Mémoires secrets*, 336.

19. Gabrielle-Pauline, comtesse d'Adhémar, *"Ma reine infortunée": Souvenirs de la comtesse d'Adhémar, dame du palais de Marie-Antoinette* (Paris: Plon, 2006), 144.

20. Weber, *Queen of Fashion*, 106.

21. Weber, 104; translation modified. As the Goncourt brothers note, imitations of the duchesse de Lauzun's pouf soon proliferated, as fashion mavens' "heads became landscapes [and] groves, with brooks running through, with shepherds, shepherdesses, and sheep appearing in them." These elements too, of course, ornament the Jardin de Monceau. See Goncourt and Goncourt, *Woman of the Eighteenth Century*, 235.

22. Although Poland was partitioned in 1772, Bertin did not introduce the *robe à la polonaise* until 1776. She developed the *circassienne* and the *turque* in 1779.

23. On the "convenience" of the *polonaise* for walks "in the country," see *La Gallerie des modes et costumes français*, vol. 1 (Paris: Esnauts et Rapilly, 1779), cahier VII, 2.

24. The balance is made up by two *robes à l'anglaise*, a modified court dress also popularized by Bertin and her clientele. Introduced toward the end of the 1770s, the *anglaise* became one of the most stylish silhouettes of the following decade.

Carmontelle's Portraits of Musicians

The author would like to thank Laurence Chatel de Brancion for attentively reading the initial draft of this contribution. It was translated from the French by Andrew Ayers and Joseph Disponzio.

1. See appendix 1 for Carmontelle's portraits of composers, musicians, and other individuals grouped by musical attributes, and appendix 2 for a comprehensive index of persons represented in the portraits.

2. F[rançois]-A[natole] Gruyer, *Chantilly: Les portraits de Carmontelle* (Paris: Plon-Nourrit, 1902).

3. Georg Kinsky, *Album musical* (Leipzig: Breitkopf & Härtel; Paris: Delagrave, 1929), 214, 278.

4. François Lesure, *Musica e Società* (Milan: Istituto Editoriale Italiano, 1966), pl. 13 (the marquis de Croimare and his daughter) and pl. 16 (M. Philippe, Mlle Delon, and M. Tellier).

5. Albert Pomme de Mirimonde, *L'Iconographie musicale sous les rois Bourbons: La musique dans les arts plastiques (XVIIᵉ-XVIIIᵉ siècles)*, vol. 2 (Paris: Éditions A. et J. Picard, 1977), 29, 31, 36, 39, 41, 44, 52, 54, 66–69, 99–100.

6. Florence Gétreau, "Carmontelle et la société musicienne de son temps," in *Instrumentistes et luthiers parisiens. XVIIᵉ-XIXᵉ siècles*, ed. Florence Gétreau, ex. cat. (Paris: Délégation à l'action artistique de la Ville de Paris, 1988), 106–7.

7. Mary Cyr, "Carmontelle's Portraits of 18th-Century Musicians," *Musical Times* 158, no. 1941 (Winter 2017), 39–54.

8. Florence Gétreau, "The Portraits of Rameau: A Methodological Approach," *Music in Art: International Journal for Music Iconography* 36, nos. 1–2 (2011), 281–82, 285–88.

9. Florence Gétreau, "Satirical Portraits and Visual Lampoons of Rameau and His Works," *Early Music* 44, no. 4 (2016), 527–29.

10. Jana Frankova, *La Migration des musiciens dans l'Europe des Lumières: Le cas de Joseph Kohaut (1734–1777)* (PhD diss., Université de Paris-Sorbonne, 2016), 2:7–9. Frankova's attribution corrects that of Gruyer, who names Charles (Karl) Kohaut, Josef's older brother, as the portrait's subject.

11. Joël Dugot, "Parcours, détours et pièges," in "De l'image à l objet: La méthode critique en iconographie musicale: In memoriam Geneviève Thibault de Chambure (1902-1975)," ed. Tilman Seebass, special issue, *Imago Musicae* 4 (1987), 251–54.

12. *Correspondance littéraire, philosophique et critique de Grimm et de Diderot depuis 1753 jusqu'en 1790*, nouv. éd., vol. 5 (Paris: Chez Furne, 1829), 445 (June 15, 1768).

13. The Concert Spirituel was a Paris-based association of musicians that gave public concerts at the Salle des Suisses in the Tuileries from 1725 to 1790.

14. [Jean-Louis-Ignace de la Serre], *Pyrame et Thisbé, tragédie. représentée par l'Académie royale de musique l'an 1726* (Paris: Chez la V. Delormel & Fils, 1759), 47.

15. Musique du Roi: The ensemble of musicians attached to the various divisions of the royal household—the *Chapelle* (chapel), *Chambre* (Chamber), and *Écurie* (stables).

16. Cyr, "Carmontelle's Portraits," 51.

17. Florence Gétreau, "Retour sur les portraits de Mozart au clavier: Un état de la question," in *Cordes et claviers au temps de Mozart; Bowed and Keyboard Instruments in the Age of Mozart*, ed. Thomas Steiner, Actes des Rencontres Internationales harmoniques, Lausanne, Switzerland, April 2006 (Bern: Peter Lang, 2010), 77–82.

18. Gruyer errs in describing this portrait. He has Jean-Aimé Vernier playing the oboe, but Vernier is not known to have ever played that instrument. Rather, the oboist is Ignace Prover, not "Philippe Provers" as he is called elsewhere by Gruyer.

19. Florence Gétreau, "The Horn in Seventeenth- and Eighteenth-Century France: Iconography Related to Performances and Musical Works," in *The Horn—History and Musical Use*, ed. Boje E. Hans Schmuhl, Monika Lustig (Michaelstein: Stiftung Kloster Michaelstein, 2006), 43–76.

20. Thomas Vernet, "Que leurs plaisirs ne finissent jamais," *Spectacles de cour: Divertissements et mécénat musical du Grand Siècle aux Lumières: L'exemple des princes de Bourbon Conti* (PhD diss., École Pratique des Hautes Études, 2010), 141 et seq.

21. Florence Gétreau and Denis Herlin, "Portraits de clavecins et de clavecinistes français. II," *Musique • Images • Instruments* 3 (1998), 64–88.

22. Gétreau and Herlin, "Portraits de clavecins," 77; Cyr, "Carmontelle's Portraits," 45–47.

23. David Hennebelle, who has meticulously gone through the available archives, does not mention her in his *Les Concerts de la Reine (1725–1768)* (Lyon: Symetrie, 2015).

24. Florence Gétreau, "Recent Research about the Voboam Family and Their Guitars," *Journal of the American Musical Instrument Society* 31 (2005), 5–66.

25. James Tyler, "Mandora," in *The New Grove Dictionary of Musical Instruments*, 2nd ed., ed. Laurence Libin (London: Oxford University Press, 2014), 3:385–87.

26. Florence Gétreau, "Les belles vielleuses au XVIIIe siècle: Du triomphe au sarcasmes," *Musique • Images • Instruments* 16 (2016), 74–109.

27. Gétreau, "The Horn in Seventeenth- and Eighteenth-Century France," 54–59.

28. Composer André Campra (1660–1744) was born in Aix-en-Provence. He became *maître de musique* in Notre-Dame de Paris in 1694. Campra devoted himself to musical theater and popularized the *tambourin de Provence*, an instrument of his native region.

29. Robert Adelson, "The Viscountess de Beaumont's Harp and Music Album (1780)," *Galpin Society Journal* 62 (2009), 160.

30. Felia Bastet, "La princesse Kinsky et la musique," in *Le Faubourg Saint-Germain. La rue Saint-Dominique. Hôtels et amateurs*, ex. cat. (Paris: Musée Rodin, 1984), 117–18, no. 163; Florence Gétreau, "Les instruments reflets de l'histoire du goût," *Musiques et musiciens au faubourg Saint-Germain*, ed. Jean Gallois, ex. cat (Paris: Délégation à l'action artistique de la ville de Paris, 1996), 56–58, "Clavecin de la princesse Kinsky."

"The Habit of Seeing the Same Things Often"

1. David L. Hays, "'This is Not a *Jardin anglais*': Carmontelle, the Jardin de Monceau, and Irregular Garden Design in Late-Eighteenth-Century-France," in *Villas and Gardens in Early Modern Italy and France*, ed. Mirka Beneš and Dianne Harris (Cambridge: Cambridge University Press, 2001), 322.

2. See Hays, this volume.

3. Thomas Blaikie, *Diary of a Scotch Gardener at the French Court at the End of the Eighteenth Century*, ed. Francis Birrell (London: George Routledge & Sons, Ltd., 1931), 134. [Ed. nt. Blaikie's spelling has been silently corrected.]

4. Blaikie, *Diary of a Scotch Gardener*, 178–9.

5. Claude-Henri Watelet, *Essay on Gardens: A Chapter in the French Picturesque*, trans. Samuel Danon (Philadelphia: University of Pennsylvania Press, 2003), 38.

6. René-Louis de Girardin, *An Essay on Landscape; or, on the Means of Improving and Embellishing the Country Round our Habitations*, trans. Daniel Malthus (London: J. Dodsley, 1783), 97–98.

7. Girardin, *An Essay on Landscape*, 98.

8. Girardin, 98–99.

9. Hays, "'This is Not a *Jardin Anglais*,'" 310.

10. David L. Hays, "Carmontelle's Design for the Jardin de Monceau: A Freemasonic Garden in Late-Eighteenth-Century France," *Eighteenth-Century Studies* 32, no. 4 (Summer 1999): 447–62.

11. Blaikie, *Diary of a Scotch Gardener*, 179.

12. Blaikie, 143.

13. Blaikie, 142.

14. Blaikie, 144.

15. Blaikie, 159.

16. Blaikie, 162.

17. Blaikie, 143.

18. Blaikie, 168–69.

19. Blaikie, 188–89.

20. Humphry Marshall. *Arbustrum Americanum: The American Grove, or, An Alphabetical Catalogue of Forest Trees and Shrubs, Natives of the American United States, arranged according to the Linnæan System* (Philadelphia: Joseph Crukshank, 1785), ix.

21. Stephen Girard to Humphry Marshall, Philadelphia, August 11, 1791, Marshall Papers, William L. Clements Library, University of Michigan.

22. Comtesse de Tessé to Humphry Marshall, December 6, 1787, Marshall Papers.

23. Louis-Philippe-Joseph, duc d'Orléans, to John Vaughan and Humphry Marshall, ca. 1785, Marshall Papers.

24. "Listes des Graines et plantes arrives d'Amerique en 1788 et semes dans les differents jardins de Monseigneur le duc d'Orléans avec la note des frais y relative," Paris, Archives Nationales, AJ 115/511, pièce 473.

Monceau, the Mémoires secrets, *and the* Colisée

1. These articles, probably initially distributed as manuscript newsletters, were gathered and reprinted by Mathieu-François Pidansat de Mairobert in 1777 as the *Mémoires secrets*. On the *Mémoires secrets*, see Robert Tate and Tawfik Mekki-Berrada, "Mathieu Pidansat de Mairobert (1727–1779)," in Anne-Marie Mercier-Faivre and Denis Reynaud, eds., *Dictionnaire des journalistes (1600–1789)*, exp. electronic. ed. (Lyon, France: UMR LIRE, Institut des Sciences de l'Homme, n.d.); orig. print ed. (Grenoble: Presses Universitaires de Grenoble, 1976), http://dictionnaire-journalistes.gazettes18e.fr/dictionnaires-presse-classique-mise-en-ligne.

2. [Mathieu-François Pidansat de Mairobert, ed.], *Mémoires secrets pour servir à l'histoire de la république des lettres en France, depuis 1762 jusqu'à nos jours* (London: John Adamson, 1784–1789), 7:15.

3. Pidansat de Mairobert, *Mémoires secrets*, 7:20.

4. Pidansat de Mairobert, *Mémoires secrets*, 7:48.

5. See Antoine-Joseph Loriot, *Mémoire sur une découverte dans l'art de bâtir faite par le sieur Loriot, mécanicien, pensionnaire du Roi; dans lequel on rend publique, par ordre de Sa Majesté, la méthode de composer un ciment ou un mortier propre à une infinité*

d'ouvrages, tant pour la construction que pour la décoration (Paris: Michel Lambert, 1774). The author thanks Basile Baudez for his insights into Loriot's invention and its significance.

6. On the accessibility of Monceau to elite visitors, see Gauthier de Simpré, *Voyage en France de M. le comte de Falckenstein* (Paris: Cailleau, 1778), 1:55.

7. Emmanuel de Croÿ-Solre, *Journal inédit du duc de Croÿ publié. d'après le ms. autographe conservé à la Bibliothèque de l'Institut* (Paris: E. Flammarion, 1906), 3: 211 (June 18, 1775).

8. Luc-Vincent Thiéry, *Guide des amateurs et des étrangers voyageurs aux environs de Paris* (Paris: Hardouin et Gattey, 1788), 246.

9. Charles-Joseph de Ligne, *Coup d'oeil sur Beloeil et sur une grande partie des jardins de l'Europe* (Leopoldberg: Walther, 1795), 2:56.

10. On exile and its evolving role in absolutist political culture, see Julian Swann, *Exile, Imprisonment, or Death: The Politics of Disgrace in Bourbon France, 1610–1789* (Oxford: Oxford University Press, 2017).

11. While Chartres held the title of *duc*, his descent from France's royal family made him a prince of the blood (*prince du sang*). This term designated members of the cadet branches of the Bourbon family. As descendants of Saint Louis who were in the line of succession, the princes of the blood occupied a middle ground between the royal family and the nobility. Chartres's father, the duc d'Orléans, was the titular First Prince of the Blood, and thus leader of this subcategory within the royal family, an honor that Chartres would inherit upon his father's death in 1785. Traditionally the princes of the blood were seen as intercessors who could advise the monarch when he was receiving bad advice from his ministers, and thus act as a safeguard for the historically constituted rights of the king's subjects and orders.

12. On the *partie patriote*, see Durand Echeverria, *The Maupeou Revolution: A Study in the History of Libertarianism in France, 1770–1774* (Baton Rouge, LA: Louisiana State University Press, 1985), and Julian Swann, *Politics and the Parlement of Paris under Louis XV, 1754–1774* (Cambridge: Cambridge University Press, 1995).

13. Pidansat de Mairobert, *Mémoires secrets*, 7:87.

14. Mathieu-François Pidansat de Mairobert and Moufle d'Angerville, *Journal historique de la Révolution opérée dans la Constitution de la monarchie françoise par M. de Maupeou, chancelier de France*, vol. 4 (London, 1774), 214.

15. On the Colisée and the informal interactions that took place there, see John Goodman, "'Altar against Altar': The Colisée, Vauxhall Utopianism and Symbolic Politics in Paris (1769–77)," *Art History* 15, no. 4 (1992): 434–69.

16. For a period account of the Colisée, see Georges-Louis Le Rouge, *Description du Colisée élevé aux Champs-Élysées sur les dessins de M. Le Camus; par le sieur Le Rouge, ingénieure géographe du Roi* (Paris: Le Rouge, 1771).

17. An article from March 31, 1769, announcing the project described Choiseul's involvement as an open secret (Pidansat de Mairobert, *Mémoires secrets*, 4:219–20).

18. "Le Camus" in Michel Gallet, *Les Architectes parisiens du XVIIIᵉ siècle : Dictionnaire biographique et critique* (Paris: Mengès, 1995), 290.

19. Articles describing the Chartres's appearance at Colisée appeared in Pidansat de Mairobert and Moufle d' Angerville, *Journal historique de la Révolution*, May 27, 1771, and July 28, 1772, and Pidansat de Mairobert, *Mémoires secrets*, December 18, 1771; these references are cited in Amédée Britsch, *La Maison d'Orléans à la fin de l'ancien régime. La Jeunesse de Philippe-Égalité (1747–1785), d'après des documents inédits* (Paris: Payot, 1926), 161.

20. David L. Hays, "Carmontelle's Design for the Jardin de Monceau: A Freemasonic Garden in Late-Eighteenth-Century France," *Eighteenth-Century Studies* 32, no. 4 (1999): 447–62.

21. On the pre-Revolutionary significance of the Phrygian or Liberty cap, see J. David Harden, "Liberty Caps and Liberty Trees," *Past & Present* 146 (February 1995): 66–102.

22. Mathieu-François Pidansat de Mairobert, *L'Espion anglois ou, Correspondance secrète entre milord All'Eye et milord All'Ear* (London: J. Adamson, 1784), 1:143.

23. Pidansat de Mairobert, *Mémoires secrets*, 7:87.

24. Chartres's naval career came to a sudden and ignominious end in autumn 1778 when it was revealed that his ineptitude at reading signals had cheated the French fleet out of the rare opportunity to deal a decisive blow to the English navy during the battle of Ushant (Britsch, *La Maison d'Orléans*, 170).

25. Britsch, *La Maison d'Orléans*, 128.

26. Munro Price, *The Perilous Crown: France between Revolutions, 1814–1848* (London: Macmillan, 2007), 207.

A Spade in the Garden or a Garden in Spades

1. On Carmontelle, see F-A. Gruyer, *Chantilly: Les Portraits de Carmontelle* (Paris, 1902); Augustin-Thierry, "Un amuseur oublié: Carmontelle, 1717–1806," *Revue des deux mondes* 82, no. 8 (April 15, 1912); F. Gaiffe, "Carmontelle, Peintre de mœurs," *Revue du dix-huitième siècle* 1 (January–March 1913). For a more recent appreciation, see Laurence Chatel de Brancion, *Carmontelle au jardins des illusions* (Château de Saint-Rémy-en-l'Eau: Éditons Monelle-Hayot, 2003).

2. Unless otherwise indicated anecdotes about Carmontelle in this essay are drawn from [Mathieu-François Pidansat de Mairobert, ed.], *Mémoires secrets pour servir à l'histoire de la république des lettres en France depuis 1762 jusqu'à nos jours* (London: John Adamson, 1784–1789) and *Correspondance littéraire, philosophique et critique par Grimm, Diderot, Raynal, Meister*, ed. Maurice Tourneux (Paris: Garnier Frères, 1877–88).

3. [Stéphanie-Félicité du Crest de Saint-Aubin, comtesse de Genlis, called] Mme de Genlis, *Proverbes et Comédies posthumes de Carmontel* [*sic*] (Paris: Chez Ladvocat, 1825), 1:ii.

4. [François-Auguste Fauveau], baron de Frénilly, *Souvenirs du Baron de Frénilly* (Paris, 1908), 7.

5. This brief account of the lives of the duc and duchesse d'Orléans is drawn from Amédée Britsch, *La Maison d'Orléans à la fin de l'ancien régime. La Jeunesse de Philippe-Égalité (1747–1785), d'après des documents inédits* (Paris: Payot, 1926).

6. Britsch, *La Masion d'Orléans*, 103–4.

7. Genlis, *Proverbes et Comédies posthumes*, 2:182–83.

8. Pidansat de Mairobert, *Mémoires secrets*, 4:290 (June 11, 1769).

9. Tourneux, *Correspondance littéraire*, 10:240 (May 1773).

10. Tourneux, *Correspondance littéraire*, 5:282–83 (May 1763).

11. Bernard Gagnebin, "La Gravure de 'La Malheureuse famille Calas'," *Gazette des beaux-arts*, 6th ser., no.92 (December 1978): 197–202; *Mémoires secrets*, 3:13 (March 28, 1766).

12. Pidansat de Mairobert, *Mémoires secrets*, 1:58 (March 26, 1762).

13. Pidansat de Mairobert, *Mémoires secrets*, 36:90 (October 7, 1787).

14. Tourneux, *Correspondance littéraire*, 11:40 (February 1775).

15. On Carmontelle's Dramatic Proverbs, see Jean-Hervé Donnard, *Le Théâtre de Carmontelle* (Paris: A. Colin, 1967).

16. The majority of his theater drawings survive in the Musée Condé at Chantilly.

17. For the influence of Molière and Diderot on the literary work of Carmontelle, see Donnard, *Le Théâtre*, chapter 3.

18. *Souvenirs du duc de Lévis*, quoted in Donnard, *Le Théâtre*, 68n4.

19. Frénilly, *Souvenirs*, 7.

20. Genlis, *Proverbes et Comédies*, ii.

21. Donnard, *Le Théâtre*, 97.

22. Dieting was a frequent concern in the ancien régime. Many famous women of the era were thin, including Sophie Arnould and Mlle Guimard.

23. Tourneux, *Correspondance littéraire*, 5:282 (May 1763).

24. Pidansat de Mairobert, *Mémoires secrets*, 14:238–40 (October 26 and 28, 1779). None of Carmontelle's other comedies was staged in his lifetime.

25. Carmontelle is the reputed and disputed author of a series of Salon critiques, known collectively as the *Coup de Patte* series, which appeared between 1779 and 1789. These critical pieces are impassioned and trenchant antiestablishment screeds, which condemn the institutionalization of the arts by the Académie Royale de peinture et de sculpture. The writing is heavily influenced by Diderot. Were Carmontelle's authorship confirmed, his character could be credited with increased depth and complexity, and he would also appear more contemptuous of the society that sustained him. For a full discussion of the pieces, see Laurence Chatel de Brancion's essay in this volume, and her note 91 above.

26. Pidansat de Mairobert, *Mémoires secrets*, 7:18 (June 28, 1773).

27. Thomas Whately, *Observations on Modern Gardening* (London, 1770), 151.

28. William Chambers, *Dissertation on Oriental Gardening: An Explanatory Discourse by Tan Chet-Qua of Quang-Chew-Fu, Gent*, 2nd ed. (London, 1773), 146.

29. Tourneux, *Correspondance littéraire*, 11:371–78 (November 1776).

30. Georges-Louis Le Rouge, *Jardins anglo-chinois, ou Détails de nouveaux jardins à la mode* (Paris: Le Rouge, 1776–89).

31. M. de Wolmar gives Saint Preux a tour of Stowe: "Les temps ainsi que les lieux y sont rassemblés avec une magnificence plus qu'humain." (The times as well as the places were gathered together here with a superhuman magnificence.) J. J. Rousseau, *La Nouvelle Héloïse*, ed. Daniel Mornet (Paris: Hachette, 1925), 3:238.

32. In the prospectus, Carmontelle provides the visit's duration.

33. Monceau is filled with other mythological iconography; for a discussion, see the author's "From Eden to the Seraglio: Louis Carogis de Carmontelle's *Jardin de Monceau* and the Persiflage of the Picturesque," in *Fragments: Architecture and the Unfinished. Essays Presented to Robin Middleton*, ed. Barry Bergdoll and Werner Oechslin (London: Thames and Hudson, 2006), 245–66.

34. Charles-Geneviève-Louis-Auguste-André-Thimothée d'Éon de Beaumont (1728–1810). After his return to France, he was banished to his home town of Tonnerre. See Pidansat de Mairobert, *Mémoires secrets*, 14:3–9 (April 1, 1779).

35. The choice of the horse chestnut, *Aesculus hippocastanum*, may have specific meaning as well. The tree bears fruit in leathery nut sacks, beset with prickles, which hang from its branches.

36. Le Rouge, *Jardin anglo-chinois*.

37. Geoges-Louis Leclerc, comte de Buffon, *Histoire naturelle, générale et particulière, avec la description du Cabinet du Roi*, vol.2 (Paris: Imprimerie Royal, 1749), 144–68.

38. The spelling of Monceau includes the following variations: Montceaux, Mousseaux, and Monceaux.

39. For the duc de Chartres's tenure as Grand Orient of French Freemasonry, see Britsch, *La Masion d'Orléans*, 229–57.

40. Officiating at a Masonic function, he wrote to Mme de Genlis: "on s'ennuie beaucoup ici." (I'm so bored here.) Quoted in Britsch, *La Masion d'Orléans*. 243.

41. On Monceau and Masonry, see David L. Hays, "Carmontelle's Design for the Jardin de Monceau: A Freemasonic Garden in Late-Eighteenth-Century France," *Eighteenth-Century Studies* 32, no. 4 (Summer 1999): 447–62.

42. "Avec la philosophie, on connoît l'action & la réaction, l'atmosphère, les propriétés de l'air, de l'eau, de la terre & du feu." [Carmontelle], *Proverbes dramatiques*, vol. 8 (Amsterdam: Chez Esprit, 1781), 62 (Proverb 95).

43. "There were few women, even in those times when morality was at its lowest ebb, who could vie with her in depravity, or who were so relentlessly and cynically indifferent to such poor shreds of reputation as they still had." (R. Douglas, *Sophie Arnould, Actress and Wit* [Paris: Charles Carrington, 1898]. 1–1). The Goncourt brothers offered praise: "Sophie est la personnification de toutes les héroïnes de la tragédie lyrique." (Sophie is the personification of all heroines of lyric tragedy.) (Edmond and Jules de Goncourt, *Sophie Arnould* [Paris: Bibliothèque-Charpentier, 1877], 91).

44. Commenting on her singing, Douglas writes: "As usual, her voice failed her." (Douglas, *Sophie Arnould*, 114).

45. Carmontelle, *Le Jardin anglais*. First published in 1937 in C. Brenner, "Le Développement du proverbe dramatique en France," *University of California Publications in Philology* 20, no. 1 (1937): 1–56.

46. The proverb also includes references to "thrusting grasses," whose sexual nature was identified by Robin Middleton in "Boullée and the Exotic," *AA Files* 19 (Spring 1990): 35–49.

En-jeux: *Viewing, Mapping, and Playing in Carmontelle's Prospectus and* Jardin de Monceau

I would like to thank Elizabeth Barlow Rogers for inviting me to participate in this project and Joseph Disponzio for serving as its invaluable managing editor.

1. George Levitine, "French Eighteenth-Century Printmaking in Search of Cultural Assertion," in *Regency to Empire, French Printmaking 1715–1814* (Minneapolis: Minneapolis Museum of Fine Arts, 1984), 10–21. Pierre Casselle, "Le Commerce des estampes à Paris dans la seconde moitié du XVIIIe siècle" (PhD diss., École Nationale de Chartes, 1976, on deposit in the BNF Estampes), 41, records that Claude-Louis Masquelier and François-Denis Née, the publishers of the *Jardin de Monceau*, had worked together on subscriptions previously. Chartres and Carmontelle subcontracted to them because, Caselle suggests, they were already known in the printing community for their luxury editions.

2. English gardens were sometimes referred to as *folies*, which suggest that the gardens were spaces for play and/or eccentricity. The financial investment necessary to build the gardens was also considered extravagant—a *folie*.

3. The difference between Carmontelle's text and a guidebook becomes evident in comparing it to Luc-Vincent Thiéry,

Guide des amateurs et des étrangers voyageurs à Paris, ou Description raisonnée de cette ville, de sa banlieue, & de tout ce qu'elles contiennent de remarquable (Paris: Chez Hardouin & Gattey, 1787), 1:64–73.

4. On the convergence of guidebooks, shopping, and tourism, see Natacha Coquery, *Tenir boutique à Paris au XVIIIe siècle: Luxe et demi luxe* (Paris: Editions du Comité des travaux historiques et scientifiques, 2011), 59–78.

5. Both David L. Hays and Joseph Disponzio have discussed the inconsistencies of Carmontelle's theory and his critique of contemporary Picturesque garden theory. This paper builds on their prescient scholarship. See Joseph Disponzio, "From Eden to Seraglio: Louis Carrogis de Carmontelle's Jardin de Monceau and the Persiflage of the Picturesque," in *Fragments: Architecture and the Unfinished. Essays presented to Robin Middleton*, ed. B. Bergdoll and W. Oechslin (New York and London: Thames and Hudson, 2006), 245–66; David L. Hays, "'This is Not a *Jardin anglais*': Carmontelle, the Jardin de Monceau, and Irregular Garden Design in Late-Eighteenth-Century France," in *Villas and Garden in Early Modern Italy and France*, ed. Mirka Beneš and Dianne Harris (Cambridge: Cambridge University Press, 2001), 294–326.

6. Disponzio, "Eden to Seraglio," 246, noted that "taken successfully and together the scenes imply a narrative and suggest an itinerary, though neither can be claimed." Carmontelle's first chapter included four pages of theory, a discussion of the plan, and the first six views. The following two chapters were divided into groups of six engravings, suggesting that each group of six could have been sold as an independent installment, like the cahiers of Le Rouge's publication.

7. My interpretation of the garden as a fluctuating "scape" is inspired by Arjun Appadurai, *Modernity at Large: Cultural Dimensions of Globalization* (Minneapolis: University of Minnesota Press, 1996). On gardens as gamescapes, see Susan Taylor-Leduc, "The Pleasures of Surprise: The Picturesque Garden in France" *Senses and Society* 10, no. 3 (2016): 361–80.

8. The broader subject of text and image in eighteenth-century book illustration is beyond the scope of this paper. A helpful introduction to contemporary reading practices is outlined by Roger Chartier and Daniel Roche, "Les pratiques urbaines de l'imprimé," in *Histoire de l'édition française*, vol. 2, *Le Livre triomphant 1660–1830* (Paris: Promodis, 1984), 280–81, 403–29.

9. François Courboin, *L'Estampe française: Graveurs et marchands*, Bibliothèque de l'art du XVIIIe siècle (Brussels et Paris: G van Oest, 1914), 95–137. For a better understanding of where engravers were located in Paris, see Hannah Williams and Chris Sparks, *Artists in Paris: Mapping the 18ᵗʰ-Century Art World*, www.artistsinparis.org (accessed September 30, 2018).

10. Kristel Smentek, "Sex, Sentiment, and Speculation: The Market for Genre Prints on the Eve of the French Revolution," in *French Genre Painting in the Eighteenth Century*, Studies in the History of Art 72 (Washington, DC: National Gallery of Art, Center for Advanced Study in the Visual Arts, 2007), 220–43.

11. David L. Hays, "Mapping and 'Natural' Garden Design in Later Eighteenth-Century France: The Example of Georges-Louis Le Rouge," *Site/Lines: A Journal of Place* 12, no. 2 (Spring 2017), 6–9.

12. Georges-Louis Le Rouge, *Jardins anglo-chinois, ou Détails de nouveaux jardins à la mode*, cahier 10 (Paris: Chez le Rouge, [1783]), plates 2 and 3.

13. The publication of prints of Monceau in the Le Rouge cahier appeared when the garden was undergoing radical transformations. By then the engraver's plates may have been available for cheaper reproductions in the competitive print market. Le Rouge famously altered plates to include them in his cahiers.

14. On Carmontelle's biography and milieu, see Laurence Chatel de Brancion, *Carmontelle au jardin des illusions* (Château de Saint-Rémy-en-l'Eau: Éditions Monellle Hayot, 2003), 37–54. Carmontelle entered the Orléans household in 1759 as reader for the duc de Chartres.

15. Daniel Roche, *A History of Everyday Things: The Birth of Consumption in France, 1600–1800* (Cambridge: Cambridge University Press, 2000; Clare Haru Crowston, *Credit, Fashion, Sex: Economies of Regard in Old Regime France* (Durham: Duke University Press, 2013), 139–94.

16. *Être et paraître, la vie aristocratique au XVIIIe siècle. Trésors cachés du musée nationale de la Renaissance*, ed. Muriel Barbier (Paris, Editions Artlys, 2015), evokes how the luxury trades fostered self-fashioning.

17. William H. Sewell Jr., "Connecting Capitalism to the French Revolution: The Parisian Promenade and the Origins of Civic Equality in Eighteenth-Century France" *Critical Historical Studies* 1, no. 1 (Spring 2014), 16. As Sewell notes, the promenade was not in itself a commercial activity, but its transformation during the eighteenth century was tied closely to two of the city's most dynamic economic sectors: fashion and commercialized leisure.

18. On the history of the Palais Royal, see *Le Palais Royal*, ex. cat., Musée Carnavalet (Paris: Réunion des Musées Nationaux, 1988); Robert M. Isherwood, *Farce and Fantasy; Popular Entertainment in Eighteenth-Century Paris* (New York and Oxford: Oxford University Press, 1986), 219–22, 248–49.

19. The duc de Chartres had already underwritten one of the first Parisian Vauxhalls, the Colisée, in 1771. This vast enterprise for popular entertainment and art exhibitions was closed by the end of the decade. For an overview of the political implications of Chartres's engagement with the Colisée, see John Goodman, "'Altar against Altar': The Colisée, Vauxhall Utopianism and Symbolic Politics in Paris (1769–77)" in *Art History* 15, no.4 (1992): 434–68, and Gabriel Wick's essay in this volume. For another interpretation of the politicization of the garden, see David L. Hays, "Carmontelle's Design for the Jardin de Monceau: A Freemasonic Garden in Late-Eighteenth-Century France," *Eighteenth-Century Studies* 32, no. 4 (Summer 1999), 447–92.

20. Barbara Maria Stafford, *Artful Science: Enlightenment Entertainment and the Eclipse of Visual Education* (Boston: MIT, 1994), 29–35, was one of the first scholars to highlight the interconnections of entertainment culture, magic, and educational pursuits.

21. See note 5 above.

22. The topic of orientalism is beyond the scope of this essay, but David L. Porter, "Monstrous Beauty: Eighteenth-Century Fashion and the Aesthetics of the Chinese Taste," in "Aesthetics and the Disciplines," special issue, *Eighteenth-Century Studies*, 35, no. 3 (Spring 2002), 395–411, provides an overview of how foreign objects stimulated a range of aesthetic issues. For an overview of the desire to record foreigners who visited France, notably at Versailles, see Meredith S. Martin, "Ambassades extraordinaires et visiteurs des contrées lointaines" in *Visiteurs de Versailles: Voyageurs, princes, ambassadeurs, 1682–1789*, ex. cat. , Château de Versailles and Metropolitan Museum of Art (Paris: Gallimard, 2017), 138–206.

23. Katie Scott, "Playing Games with Otherness: Watteau's Chinese Cabinet at the Château de la Muette," *Journal of the Warburg and Courtauld Institutes* 66 (2003), 189–248, has argued that jokes and play informed rococo viewing practices of decorative interiors.

24. The history of jokes and enigmas would have been familiar in Orléans court circles. A pattern of ridicule had been well established since the reign of Louis XIV, as explained by Betsy Rosasco, "Masquerade and Enigma at the Court of Louis XIV," in "Images of Rule: Issues of Interpretation," special issue, *Art Journal* 48, no. 2 (Summer 1989), 144–49.

25. Charles-François-Nicolas Le Maître de Claville, *Traité du vrai mérite de l'homme considéré dans tous les âges et dans toutes les conditions, avec des principes d'éducation propres à former les jeunes gens à la vertu*, 2nd ed. (Paris: Savoye, 1776), 277.

26. On Chartres's obsessive gambling, see Amédée Britsch, *La Maison d'Orléans à la fin de l'ancien régime. La Jeunesse de Philippe-Égalité (1747–1785), d'après des documents inédits* (Paris: Payot, 1926). On gambling, see Thomas Kavanagh, *Enlightenment and the Shadows of Chance* (Baltimore: Johns Hopkins University Press, 1993); Eve Netchine, ed., *Jeux de princes, jeux de vilains* (Paris: Bibliothèque Nationale de France and Éditions du Seuil, 2009).

27. Manfred Zollinger, "Homo Ludens, Homo Legens," in Netchine, *Jeux de princes*, 68–72.

28. Thierry Depaulis, "Temps nouveau, jeux nouveaux," in Netchine, *Jeux de princes*, 39–65.

29. Jean Dusaulx, *Lettres et réflexions sur la fureur du jeu, auxquelles on a joint une autre lettre morale* (Paris: Lacombe, 1775).

30. Disponzio, "Eden to Seraglio," 246, commented that it was not possible to correlate the plan and plates.

31. Michel Conan, "Friendship and Imagination in French Baroque Gardens before 1661," in *Baroque Garden Culture: Emulation, Sublimation, Subversion*, ed. Michel Conan, Dumbarton Oaks Colloquium on the History of Landscape Architecture 25 (Washington, DC: Dumbarton Oaks, 2005), 323–83, and Conan, "The Conundrum of La Nôtre's Labyrinth," in *Garden History: Issues, Approaches, Methods*, ed. John Dixon Hunt, Dumbarton Oaks Colloquium on the History of Landscape Architecture 13 (Washington, DC: Dumbarton Oaks, 1992), 131–36; Franz Reitinger, "Mapping Relationships: Allegory, Gender and the Cartographical Image in Eighteenth-Century France and England," *Imago Mundi* 51 (1999): 106–30.

32. Jean Marie Lhôte, *Histoire des jeux de société* (Paris: Flammarion, 1994), 255–62.

33. Hays, "This is Not a *Jardin anglais*," 296, points out that Genlis and Carmontelle were close friends. Genlis wrote a posthumous introduction to Carmontelle's proverbs and praised his ability to capture society's mores.

34. Barbier, *Être et paraître*, 72–82, provides a brief overview of cards and game pieces.

35. Mimi Hellman, "Interior Motives: Seduction by Decoration in Eighteenth-Century France," in *Dangerous Liaisons: Fashion and Furniture in the Eighteenth Century*, ed. Harold Koda and Andrew Bolton (New York: Metropolitan Museum of Art, 2006), 18–19. Hellman persuasively argues that the elite social body was dedicated to pleasure and visual seduction.

36. On the significance of vertigo in painted representations of gaming, see Jennifer Milam, *Fragonard's Playful Paintings: Visual Games in Rococo Art*, Critical Perspectives in Art History 2 (Manchester, UK: Manchester University Press, 2007).

37. Gilles-Antoine Langlois, *Folies, tivolis et attractions: Les premiers parcs de loisirs parisiens* (Paris: Délégation de l'Action Artistique de la Ville de Paris, 1991).

38. Langlois, 22–36, 138–48.

Image Credits

FRONTISPIECE: Carmontelle, *Monsieur de Carmontelle, Reader of the Duc d'Orleans*, 1762. Black lead, gouache, sanguine, and watercolor on paper, 9 13/16 x 7 1/2 in. (25 x 19 cm). Chantilly, Musée Condé, inv. no. Car52 (© RMN-Grand Palais/Art Resource, NY; photo: René-Gabriel Ojéda).

PLATES, PAGES 26–61, are from Carmontelle's *Jardin de Monceau* in the collection of the Oak Spring Garden Foundation, Upperville, VA.

PAGE 63. Carmontelle, *Figures Walking in a Parkland*, ca. 1783–1800 (detail). Watercolor and gouache, with traces of black chalk underdrawing, on Whatman translucent vellum, 18 5/8 x 148 7/16 in. (47.3 x 337 cm). J. Paul Getty Museum, 96.GC.20.

PAGE 95, above: Carmontelle, *Vers l'île enchantée*, ca. 1785–90 (detail). Watercolor and gouache on Whatman translucent vellum, 16 9/16 x 25 9/16 in. (42 x 70.5 cm). Private collection.

PAGE 95, below: Carmontelle, *Les Sorbets sous le parasol*, ca.1785–90 (detail). Watercolor and gouache on Whatman translucent vellum, 16 9/16 x 25 9/16 in. (42 x 65 cm). Private collection.

1. Carmontelle, *Portrait of Louis-Philippe, Duc d'Orléans, and His Son Louis-Philippe-Joseph, Duc de Chartres*, 1759. Etching: sheet, 11 11/16 × 7 7/8 in (29.7 x 20 cm); image, 11 5/8 × 7 3/4 in. (29.5 x 19.7 cm). New York, Metropolitan Museum of Art, The Elisha Whittlesey Collection, The Elisha Whittlesey Fund, 1996, 1996.275.

2. Carmontelle, *La Société du Palais-Royal*, ca. 1773–75. Red and black chalk, watercolor, and gouache with white heightening on paper, 12 11/16 x 16 13/16 in. (32.2 x 42.7 cm). Sale, Sotheby's Paris, September 29–30, 2015, no. 7 (photo: Sotheby's).

3. Carmontelle, *Les Gentilshommes du duc d'Orléans en habit de Saint-Cloud*. Red and black chalk, watercolor, and gouache with white heightening on paper, 10 3/8 x 15 3/4 in. (26.3 x 40 cm). Sale, Sotheby's Paris, September 29–30, 2015, no. 6 (photo: Sotheby's).

4. Jean-Baptiste-Joseph Delafosse (1721–1806) after Carmontelle, *Portrait of Louis-Philippe, Duc d'Orléans, on Horseback*. Etching and engraving: sheet, 17 5/16 x 13 1/16 in. (44 x 33.1 cm); image, 16 7/8 x 10 1/2 in. (42.9 x 26.7 cm). New York, Metropolitan Museum of Art, The Elisha Whittlesey Collection, The Elisha Whittlesey Fund, 1951, 51.568.9.

5. Carmontelle, *Monsieur de Mornay, Gouverneur de Saint-Cloud*. Red and black chalk, watercolor, and gouache with white heightening on paper, 10 11/16 x 7 1/2 in. (27.1 x 19 cm). Sale, Sotheby's Paris, September 29–30, 2015, no. 10 (photo: Sotheby's).

6. Carmontelle, *The Duchess of Chaulnes as a Gardener in an Allée*, 1771. Watercolor and gouache over black and red chalk on off-white paper, 12 12 x 7 1/2 in. (31.8 x 19.1 cm). J. Paul Getty Museum, 94.GC.41.

7. Jean-Baptiste-Joseph Delafosse (1721–1806), after Carmontelle, *Leopold Mozart and His Children Maria Anna and Wolfgang Giving a Concert in Paris*, 1764. Etching and engraving: sheet (trimmed), 15 3/16 x 9 1/8 in. (38.5 x 23.1 cm). New York, Metropolitan Museum of Art, Harris Brisbane Dick Fund, 1917, 17.3.756–1354.

8. François-Denis Nee (1732–1817) after Carmontelle, *Benjamin Franklin (1706–1790)*, 1781. Engraving: image, 12 3/16 x 7 9/16 in. (31 x 19.2 cm). Washington, DC, National Portrait Gallery, Smithsonian Institution, NPG.77.218.

9. Carmontelle, *Les Comédiens en route vers le château*, ca. 1785–90 (detail). Watercolor and gouache on Whatman translucent vellum, 16 9/16 x 25 9/16 in. (42 x 70.5 cm). Private collection.

10. Carmontelle, *Le Départ de Bathilde d'Orléans, duchesse de Bourbon*, ca. 1785–90 (detail). Watercolor and gouache on Whatman translucent vellum, 16 9/16 x 25 9/16 in. (42 x 70.5 cm). Private collection.

11. Carmontelle, *Le Duc de Chartres en habit de Saint-Cloud*. Red and black chalk, watercolor, and gouache on paper, 11 5/16 x 7 in. (28.7 x 17.8 cm). Sale, Sotheby's Paris, September 29–30, 2015, no. 8 (photo: Sotheby's).

12. Jean-Baptiste Pigalle (1714–84), *Madame de Pompadour en Amitié*, 1753. Marble, 65 3/8 x 24 7/16 x 21 5/8 in. (166 x 62 x 55 cm). Paris, Musée du Louvre, Gift of Guy de Rothshild, 1974, RF3026 (© RMN-Grand Palais; photo: Adrien Didierjean).

13. Jean-Antoine Houdon (1741–1828), *Bather* (from a fountain group), 1782. Marble, 47 x 43 x 28 in. (119.4 x 109.2 x 71.1 cm). New York, Metropolitan Museum of Art, Bequest of Benjamin Altman, 1913, 14.40.673.

14. Edme Bouchardon (1698–1762), *Faune endormi*, 1730. Marble, 72 7/16 x 56 1/8 x 47 1/16 in. (184 x 142.5 x 119.5 cm). Paris, Musée du Louvre, MR1921 (© RMN-Grand Palais; photo: René-Gabriel Ojéda).

15. Carmontelle, detail from "Mémoire sur les tableaux transparents du citoyen Carmontelle, l'an 3è de la Liberté" (1794).

Watercolor and gouache. Paris, Bibliothèque de l'Institut national d'histoire de l'art, Autographes 008.1. ©Bibliothèque numérique de l'INHA.

16. Carmontelle, *The Farm*, ca. 1785–90 (detail). Watercolor and gouache on Whatman translucent vellum, 16 9/16 x 25 9/16 in. (42 x 65 cm). Private collection.

17. Carmontelle, illustration from "Mémoire sur les tableaux transparents du citoyen Carmontelle, l'an 3è de la Liberté" (1794). Watercolor and gouache, 8 7/10 x 7 ½ in (22.2 x 19.0 cm). Paris, Bibliothèque de l'Institut national d'histoire de l'art, Autographes 008.1. ©Bibliothèque numérique de l'INHA.

18. Carmontelle, illustration from "La Perspective démontrée à l'usage des jeunes gens qui savent la géométrie et le dessin," 1794–95. Paris, Bibliothèque de l'Institut national d'histoire de l'art, Ms 845. ©Bibliothèque numérique de l'INHA.

19. Circular Temple from Monceau on the Île de la Jatte, which was once part of the park of the Château de Neuilly (photo: Gabriel Wick).

20. The Naumachia in Parc Monceau (photo: Gabriel Wick).

21. The Pyramid in Parc Monceau (photo: Gabriel Wick).

22. *The Statue of Liberty, near the Parc Monceau*. From *Le Monde Illustré*, July 12, 1884.

23. Jean Delagrive (1689–1757), *Carte topographique des environs et du plan de Paris levée par M. l'Abbé Delagrive et copiée selon l'original parisien par les héritiers de feu Dr Homann* (Nuremberg: Les Héritiers d'Homann, 1735) (photo: gallica.bnf.fr / Bibliothèque nationale de France).

24. Louis-Marie Colignon (d. 1794), "Plan général" from "Vente à vie de 3 arpents 1/2 par Marie Louis Colignon, architecte et entrepreneur des bâtiments du roi, rue du Four, à Louis-Philippe-Joseph d'Orléans, duc de Chartres, d'un terrain près de la barrière" (December 19, 1769). Paris, Archives Nationales, MC/RS//1078 (former MC/ET/LIII/459).

25. Site formation at the Jardin de Monceau between 1769 and 1776. Drawing by David L. Hays.

26. Joseph-Jacques Blessebois de la Garenne, "Terrein sis au Terroir de Clichy," July 12, 1773. Paris, Bibliothèque nationale de France, Estampes, Reserve VE-53 (I)-FT 5, Collection Destailleur 928 (photo: gallica.bnf.fr / Bibliothèque nationale de France).

27. Detail of 26.

28. Roussel, *Paris, ses fauxbourgs et ses environs où se trouve le détail des villages, châteaux, grands chemins pavez et autres, des hauteurs, bois, vignes, terres et prez, levez géométriquement* (Paris: Jaillot, 1730–39). Washington, DC, Library of Congress, Geography and Map Division, G5834.P3 1731.R6.

29. Detail of plate 1 from Carmontelle's *Jardin de Monceau* (yellow line added).

30. Anonymous, "Plan d'un Jardin à Mouceau" (ca. 1788). Paris, Bibliothèque de l'Institut national d'histoire de l'art, collections Jacques-Doucet, F° I 103.

31. Carmontelle, *Madame la duchesse de Chartres*, 1770. Watercolor, black lead, gouache, and sanguine, 7 x 11 7/8 in (18 x 30 cm). Chantilly, Musée Condé, Estate of Henri d'Orléans, duc d'Aumale, no. Car4 (© RMN-Grand Palais/Art Resource, NY; photo: René-Gabriel Ojéda).

32. Artist unknown (French, 18th century), *Coëffure au Que-za-co vue par derrière*, 1778. Hand-colored engraving on laid paper, approx. 2 1/3 x 1 2/3 in. (6.0 cm x 4.1 cm). Boston, Museum of Fine Arts, Elizabeth Day McCormick Collection; gift to the Museum of Fine Arts, 1944, 44.1258 (photo: MFA, Boston).

33. After Pierre-Thomas LeClerc (ca. 1740–after 1799), *Jeune Dame en Circassienne de gaze d'Italie . . . coëffée [sic] d'un pouf au fichu garni de perles avec une plume sur le côte gauche, à la mode Asiatique*. Hand-colored engraving on laid paper, 7 x 4 1/3 in. (17.7 cm x 11.1 cm). Amsterdam, Rijksmuseum, purchased with the support of the F. G. Waller Fonds, September 30, 2009, RP-P-2009-1226 (photo: Rijksmuseum).

34. Artist unknown (French, 18th century), *Bonnet au pouf*, 1776. Hand-colored engraving on laid paper, approx. 2 1/3 x 1 2/3 in. (6.0 cm x 4.1 cm). Amsterdam, Rijksmuseum, purchased with the support of the F. G. Waller Fonds, September 30, 2009, RP-P-2009-1136 (photo: Rijksmuseum).

35. Designer unknown (French, 18th century), *Robe à la polonaise*, ca.1780. Hand-embroidered silk. New York, Metropolitan Museum of Art, purchased with a gift from Mr. and Mrs. D1976.146a, b (photo: Metropolitan Museum of Art).

36. Nicolas Dupin (active 1776–87) after Claude-Louis Desrais (1746–1816), *Circassienne de taffetas à bandes de ruban avec la juppe d'une autre couleur garnie de gaze*, 1779. Hand-colored engraving on laid paper, 12 1/4 x 10 1/4 in. (31 cm x 26 cm). Boston, Museum of Fine Arts, Elizabeth Day McCormick Collection; gift to the MFA, Boston, 1944, 44.1405 (photo: MFA, Boston).

37. Nicolas Dupin (active 1776–87) after Claude-Louis Desrais (1746–1816), *Demoiselle à la promenade du matin, en Polonoise garnie en tuyaux*, 1778. Hand-colored engraving on laid paper, 12 1/4 x 10 1/4 in. (31 cm x 26 cm). Boston, Museum of Fine Arts, Elizabeth Day McCormick Collection; gift to the MFA, Boston, 1944, 44.1318 (photo: MFA, Boston).

38. Carmontelle, *Monsieur de Kohault, musicien autrichien*, 1764. Gouache, graphite, red chalk, and watercolor on paper, 11 13/16 x 6 7/8 in. (30 x 17.5 cm). Chantilly, Musée Condé, Car426 (© RMN-Grand Palais; photo: René-Gabriel Ojéda).

39. Carmontelle, *Madame de Maupassant, femme d'un commissaire des guerres*, 1759. Gouache, graphite, red chalk, and watercolor on paper, 10 5/8 x 6 11/16 in. (27 x 17 cm). Chantilly, Musée Condé, Car294 (© RMN-Grand Palais; photo: René-Gabriel Ojéda).

40. Carmontelle, *Madame de Montainville*, 1758. Watercolor, graphite, and chalk on paper, 10 1/16 x 6 15/16 in. (25.6 x 17.7 cm). Amsterdam, Rijksmuseum, Rijksprentenkabinet, RP-T-1961-18 (photo: Rijksmuseum).

41. Carmontelle, *Monsieur and Madame Blizet with Monsieur Le Roy the Actor*, ca. 1765. Watercolor and gouache over red and black chalks with touches of graphite on laid paper, 10 15/16 x 7 5/16 in. (27.8 x 18.5 cm). Washington, DC, National Gallery of Art, Gift of Ruth Carter Stevenson in memory of Sir Geoffrey Agnew, 1987.56.1 (photo: ©National Gallery of Art).

42. Carmontelle, *Narcisse, nègre du duc d'Orléans*. Gouache, graphite, red chalk, and watercolor on paper, 12 5/8 x 7 1/16 in. (32 x 18 cm). Chantilly, Musée Condé, Car85 (© RMN-Grand Palais; photo: René-Gabriel Ojéda).

43. Attributed to Carmontelle, *Vue des jardins de Monceau*, ca. 1778. Oil on canvas, 25 5/8 x 36 3/4 in. (65 x 93.5 cm). Paris, Musée Carnavalet, CARP1784 (photo: Bridgeman Images).

44. Hercules Segers (1589/90–ca.1640), *The Large Tree*, ca. 1628–1629. Etching and black chalk on white paper, 8 1/2 x 10 7/8 in. (21.7 x 27.7 cm) (sheet). Amsterdam, Rijksmuseum, transferred from the Koninklijke Bibliotheek, collection of Pieter Cornelis Baron van Leyden, 1816, RP-P-OB-849 (photo: Rijksmuseum).

45. Louis de Chastillon (1639–1734), *Plante chinoise (aubépine)*, 1676. Etching and engraving, 16 1/8 in x 11 3/4 in (41.1 cm x 29.9 cm). Paris, Musée du Louvre, Atelier de la chalcographie, 5500 C/ Recto (© RMN-Grand Palais).

46. Georges-Louis Le Rouge (ca. 1713–1780), *Coupe du nouveau Berceau ou, Jardin d'Hiver de S.A.S.M. le Duc de Chartres à Mousseau* (*Jardins anglo-chinois*, cahier 10, plate 3), [1783]. Paris, Bibliothèque de l'Institut national d'histoire de l'art, collections Jacques Doucet, NUM 4 RES 216 (10).

47. Georges-Louis Le Rouge (ca. 1713–1780), *Coupe du Jardin d'Hiver* (detail of plate 2 from *Jardins anglo-chinois*, cahier 10), 1783. Paris, Bibliothèque de l'Institut national d'histoire de l'art, collections Jacques Doucet, NUM 4 RES 216 (10).

48. Gabriel-Jacques de Saint-Aubin (1724–1780), *La Fête chinoise au Colisée*, 1772. Pen and black ink, gouache, watercolor, graphite, 4 1/8 x 8 3/4 in. (10.5 x 22.5 cm). Paris, Musée du Louvre, NV32751 (© RMN-Grand Palais/Art Resource, NY; photo: Jean-Gilles Berizzi).

49. Gabriel-Jacques de Saint-Aubin (1724–1780), *Nocturnal Water Spectacle in the Colisée*, 1772. Ink, gouache, watercolor, and graphite on paper, 6 1/2 x 9 in. (16.7 x 22.7 cm). Rotterdam, Museum Boijmans Van Beuningen.

50. Gabriel-Jacques de Saint-Aubin (1724–1780), *The Naumachia of Monceau*, 1778. Black chalk, ink, watercolor, and body color on paper, 31 x 40 in. (79 x 102 cm). Location unknown (former collection Mme H. Dacier).

51. Nicolas-Bernard Lépicié (1735–1784), *Louis-Philippe, Duke of Valois, in His Cradle*, 1774. Oil on canvas, laid down on wood panel, 21 1/2 x 16 1/8 in. (54.5 x 41 cm). Musée national du Château de Versailles et de Trianon, V2015.30 (© RMN-Grand Palais/Art Resource, NY; photo: Christohe Fouin).

52. Joseph-Siffred Duplessis (1725–1802), *The Duchess of Chartres in the Presence of the Vessel Saint-Esprit, which takes the Duke of Chartres to the Battle of Ushant*, 1777–78. Oil on canvas, 38 1/4 × 51 1/4 in. (97 × 130 cm). Chantilly, Musée Condé, PE386 (© RMN-Grand Palais/Art Resource, NY; photo: René-Gabriel Ojéda).

53. *Animaux spermatiques suivant la dernière édition de Leeuwenoek*. From Georges-Louis Leclerc, comte de Buffon, *Histoire naturelle générale et particulière avec la description du Cabinet du Roi*, vol. 2 (Paris: Imprimerie Royale, 1749), 248 (plate 7).

54. Artist unknown, *Carte et jeu allégorique du bonheur*, ca. 1780–1800. Engraving, 20 7/8 x 16 1/2 in. (53 x 42 cm). Paris, Bibliothèque Nationale de France, Département des Estampes et de la photographie.

55. Artist unknown, *Jeu de l'oie, Jeu instructif des fleurs*. Engraving, 20 7/8 x 16 1/2 in. (53 x 42 cm). Paris, Bibliothèque nationale de France, Département des Estampes et de la photographie.

56. Bordier, after a survey by Gauche, *Plan des Châteaux et Jardins de Choisi le Roy 1783* (from G.-L. Le Rouge, *Jardins anglo-chinois*, cahier 10). Engraving, 25 3/16 x 18 7/8 in. (64 x 48cm). Paris, Bibliothèque nationale de France, Département des Estampes et de la photographie.

Contributors

ANDREW AYERS studied architectural history at the Bartlett School of Architecture, University College London, and arts criticism at City University, London. A resident of France for over twenty years, he is the author of *The Architecture of Paris*, a guide to the city's built fabric from Gallo-Roman times to the present, and an architectural journalist contributing regularly to *L'Architecture d'aujourd'hui*, *The Architectural Review*, *Architectural Record*, *Icon*, *Metropolos* and *PIN-UP* (of which he is associate editor). He teaches and lectures for Paris based American universities including Columbia University and the University of Pennsylvania. In addition, he is the chief docent at the Maison de Verre, Paris. His translated works from French to English range from the writings of the Comte de Lacépède to Jean Nouvel.

LAURENCE CHATEL DE BRANCION is an award winning writer on French history and recipient of the silver medal for biography by the Académie française. She is especially known for her scholarship on Carmontelle, having written extensively on his life and work including *Carmontelle au jardins des illusions* (2003), and *Le Cinema au siècles des Lumières* (2007, published in English as *Carmontelle's Landscape Tranparencies: Cinema of the Enlightenment*). This latter work treats Carmontelle's transparencies and his experiments with animation. She has helped in the creation of a facsimile of one of Carmontelle's "films" now on display at the Musée Diderot at Langres. In 2019 she authenticated a hitherto lost manuscript by Carmontelle, his *Traité de perspective*. In addition to her work on Carmontelle she has written on the art of gardening, most recently on Hubert Robert and the garden he created for the marquis de Lafayette. Her curatorial projects include exhibitions on Lafayette, Napoleon's coronation and Benjamin Franklin. She holds a Ph.D. from Paris-Sorbonne University.

JOSEPH DISPONZIO is a preservation landscape architect with the New York City Department of Parks and Recreation. Prior to his present post he taught at the Harvard Graduate School of Design, Bryn Mawr College, and the University of Georgia. He founded and directed Columbia University's master-of-science program in landscape design. He is a noted author and scholar of garden history, specializing in the Picturesque garden tradition and the history of the profession of landscape architecture. His most recent publication is the introduction to Jean-Marie Morel's *Theory of Gardens* (Dumbarton Oaks, 2019).

FLORENCE GÉTREAU, musicologist and art historian, is director emeritus of research at the CNRS Institut de recherche en musicologie in Paris. She is the author and editor of numerous publications on organology, the sociology of music, musical iconography, and art history. Her last book, *Voir la Musique* (2017) received the Prix du Cercle Montherlant de l'Académie des Beaux-Arts and the Claire Brook Award in New York. She is the founder and director of the annual journal *Musique · Images · Instruments* and has curated many exhibitions, most recently Wine and Music: Harmony and Dissonance (Bordeaux, Cité du Vin, 2018). Past president of the Société française de musicologie (2011–15), she is a member of the directorium of the International Musicological Society.

DAVID L. HAYS is the Brenton H. and Jean B. Wadsworth Head of the Department of Landscape Architecture at the University of Illinois at Urbana-Champaign. His scholarly research explores the history of garden and landscape design in early modern Europe, contemporary landscape theory and practice, interfaces between architecture and landscape, and pedagogies of history and design. Hays is coeditor of *Forty-Five: A Journal of Outside Research* and editor of *Landscape within Architecture* (2004) and *(Non-) Essential Knowledge for (New) Architecture* (2013) (special issues of 306090 published by Princeton Architectural Press). His essays have appeared in a wide range of academic journals, including *Harvard Design Magazine*, *Eighteenth-Century Studies*, and *Polysèmes*, and as chapters in numerous scholarly books. He holds an M. Arch. from Princeton and a Ph.D. in the history of art from Yale.

ELIZABETH HYDE is professor and chair of the Department of History at Kean University. Her first book, *Cultivated Power: Flowers, Culture, and Politics in the Reign of Louis XIV* (2005) explores the collection, cultivation, and political importance of flowers in early modern France. She also edited and contributed to *A Cultural History of Gardens in the Renaissance, 1400–1650* (2013). Hyde is currently writing *Of Monarchical Climates and Republican Soil: Nature, Nation, and Botanical Diplomacy in the Franco-American Atlantic World*, a book that explores the late-eighteenth-century mission of French botanist André Michaux to study and collect North American trees. In 2017 she was a Visiting Mellon Scholar at the New York Botanical Garden. She is currently a co-recipient of a three-year Humanities Initiative Grant from the National Endowment for the Humanities to develop a history lab at her institution entitled "MakeHISTORY@Kean: William Livingston's World."

SUSAN TAYLOR-LEDUC is an independent scholar affiliated with the Centre des Recherches de Versailles. Based in Paris since 1992 she has worked as a professor, curator, critic, and university administrator. She is a noted scholar and author on the French Picturesque. Her book, *Designing Legacy: The Picturesque Garden in France 1775-1867*, is forthcoming from Amsterdam University Press (2021). She earned both her masters and doctoral degrees in the history of art from the University of Pennsylvania.

CAROLINE WEBER is a professor of French and comparative literature at Barnard College, Columbia University. Her publications include *Terror and Its Discontents* (2003) and *Queen of Fashion: What Marie Antoinette Wore to the Revolution* (2006). Her most recent book, *Proust's Duchess: How Three Celebrated Women Captured the Imagination of Fin-de-Siecle Paris* (2018), was a finalist for the Pulitzer Prize.

GABRIEL WICK lectures on the history of design and the built environment at the Paris campuses of New York University and The New School/Parsons School of Design. He received his doctorate in history from Queen Mary University of London and holds a master's in landscape architecture from the University of California Berkeley and a master's in conservation from the École nationale supérieure d'architecture de Versailles. He is the author of *Un Paysage des Lumières: Le jardin anglais du château de La Roche-Guyon* (2014), and *Le Domaine de Méréville: Renaissance d'un jardin* (2018). He co-edited *Une maison de plaisance au XVIIIème siècle: L'hôtel de Noailles à Saint-Germain-en-Laye* (2016) and *Hubert Robert et la fabrique des jardins* (2017). He curated the 2017 exhibition *Hubert Robert et la fabrique des jardins* (Château de La Roche-Guyon) and is co-curating the 2020 exhibition *Parcs des lumières, jardins de Voltaire* (Château de Ferney / CMN).

Set in the types of Pierre-Simon Fournier.
Design & typography by
Jerry Kelly.